NEUROMETHODS

Series Editor
Wolfgang Walz
University of Saskatchewan
Saskatoon, Canada

For further volumes:
http://www.springer.com/series/7657

Neuromethods publishes cutting-edge methods and protocols in all areas of neuroscience as well as translational neurological and mental research. Each volume in the series offers tested laboratory protocols, step-by-step methods for reproducible lab experiments and addresses methodological controversies and pitfalls in order to aid neuroscientists in experimentation. *Neuromethods* focuses on traditional and emerging topics with wide-ranging implications to brain function, such as electrophysiology, neuroimaging, behavioral analysis, genomics, neurodegeneration, translational research and clinical trials. *Neuromethods* provides investigators and trainees with highly useful compendiums of key strategies and approaches for successful research in animal and human brain function including translational "bench to bedside" approaches to mental and neurological diseases.

Translational Research Methods for Pediatric Brain Tumors

Edited by

Michael Goldstein

Radiation Oncology, Johns Hopkins School of Medicine, Baltimore, MD, USA

Humana Press

Editor
Michael Goldstein
Radiation Oncology
Johns Hopkins School of Medicine
Baltimore, MD, USA

ISSN 0893-2336 ISSN 1940-6045 (electronic)
Neuromethods
ISBN 978-1-0716-4753-0 ISBN 978-1-0716-4754-7 (eBook)
https://doi.org/10.1007/978-1-0716-4754-7

This Humana imprint is published by the registered company Springer Science+Business Media, LLC, part of Springer Nature.
The registered company address is: 1 New York Plaza, New York, NY 10004, U.S.A.

Preface to the Series

Experimental life sciences have two basic foundations: concepts and tools. The Neuromethods series focuses on the tools and techniques unique to the investigation of the nervous system and excitable cells. It will not, however, shortchange the concept side of things as care has been taken to integrate these tools within the context of the concepts and questions under investigation. In this way, the series is unique in that it not only collects protocols but also includes theoretical background information and critiques which led to the methods and their development. Thus, it gives the reader a better understanding of the origin of the techniques and their potential future development. The Neuromethods publishing program strikes a balance between recent and exciting developments like those concerning new animal models of disease, imaging, in vivo methods, and more established techniques, including, for example, immunocytochemistry and electrophysiological technologies. New trainees in neurosciences still need a sound footing in these older methods in order to apply a critical approach to their results.

Under the guidance of its founders, Alan Boulton and Glen Baker, the Neuromethods series has been a success since its first volume published through Humana Press in 1985. The series continues to flourish through many changes over the years. It is now published under the umbrella of Springer Protocols. While methods involving brain research have changed a lot since the series started, the publishing environment and technology have changed even more radically. Neuromethods has the distinct layout and style of the Springer Protocols program, designed specifically for readability and ease of reference in a laboratory setting.

The careful application of methods is potentially the most important step in the process of scientific inquiry. In the past, new methodologies led the way in developing new disciplines in the biological and medical sciences. For example, Physiology emerged out of Anatomy in the nineteenth century by harnessing new methods based on the newly discovered phenomenon of electricity. Nowadays, the relationships between disciplines and methods are more complex. Methods are now widely shared between disciplines and research areas. New developments in electronic publishing make it possible for scientists that encounter new methods to quickly find sources of information electronically. The design of individual volumes and chapters in this series takes this new access technology into account. Springer Protocols makes it possible to download single protocols separately. In addition, Springer makes its print-on-demand technology available globally. A print copy can therefore be acquired quickly and for a competitive price anywhere in the world.

Saskatoon, Canada ***Wolfgang Walz***

Preface

The field of pediatric neuro-oncology has experienced dramatic advances over the past two decades owing to the progress in genetic, epigenetic, and biochemical investigation of pediatric brain tumors. Application of novel research approaches has allowed us to understand the molecular drivers of carcinogenesis and identify the developmental origins of pediatric brain malignancies. Genetic mutations, epigenetic alterations, and novel signaling pathways involving oncogenic histones have been identified as crucial factors in the formation of pediatric brain tumors demonstrating the uniqueness of these malignancies within the realm of oncology.

Comprehensive genomic, epigenomic, and proteomic analyses have provided critical insights into the biology of pediatric brain cancers. Importantly, these investigations have demonstrated a substantial molecular diversity within histologically defined tumor types suggesting that a former tumor classification solely based on histology is obsolete. Instead, multiple molecular subgroups affecting patient survival and treatment response have been identified within the same tumor types such as medulloblastoma and ependymoma.

A growing role of molecular profiling in diagnostics and treatment of pediatric brain tumors indicates a need for comprehensive research models reflecting the molecular features and clinical phenotypes of these malignancies. This book focuses on translational models of pediatric brain cancers that are critical for the mechanistic investigation of tumor formation, tumor progression, and response to treatment. These models are particularly important for drug screens and preclinical testing of novel treatment approaches involving chemotherapy, radiation, and immunotherapy. Furthermore, this book illustrates the diverse methods and cutting-edge technologies used to investigate the molecular backbone of pediatric brain tumors. These novel diagnostic tools include tumor characterization from liquid biopsy, capturing the local heterogeneity of a single tumor through spatial genomics, and deciphering the unique epigenetic traits of pediatric brain tumors by profiling tumor methylome and post-translational histone modifications. Finally, the book describes experimental tools designed to explore new therapeutic strategies and optimize current treatments. Thus, it covers all critical facets of experimental pediatric neuro-oncology that will be shaping the future of the field.

Baltimore, MD, USA ***Michael Goldstein***

Contents

Contributors

MUNJAL M. ACHARYA • *Departments of Radiation Oncology, Anatomy and Neurobiology, University of California, Irvine School of Medicine, Irvine, CA, USA*

NAGI G. AYAD • *Department of Oncology, Lombardi Comprehensive Cancer Center, Georgetown University Medical School, Washington, DC, USA; Laboratory of Developmental Neurobiology, The Rockefeller University, New York, NY, USA*

SHREYA BANERJEE • *Department of Ophthalmology, Johns Hopkins University School of Medicine, Baltimore, MD, USA*

ANNELIESE CEISEL • *Department of Ophthalmology, Johns Hopkins University School of Medicine, Baltimore, MD, USA*

MILAN G. CHHEDA • *Division of Oncology, Department of Medicine, Washington University School of Medicine, St. Louis, MO, USA; BJCIH, Washington University School of Medicine, St. Louis, MO, USA*

EIRINI CHRISTODOULOU • *Children's Hospital Los Angeles, Los Angeles, CA, USA; Department of Pathology, Harvard Medical School, Boston, United States*

KALEEM COLEMAN • *The Institute for Biomedical Sciences, The George Washington University, Washington, DC, USA*

CHARLES G. EBERHART • *Department of Ophthalmology, Johns Hopkins University School of Medicine, Baltimore, MD, USA; Department of Pathology, Johns Hopkins University School of Medicine, Baltimore, MD, USA*

ISABELLE EGAN • *Department of Oncology, Lombardi Comprehensive Cancer Center, Georgetown University Medical School, Washington, DC, USA*

YIN FANG • *Laboratory of Developmental Neurobiology, The Rockefeller University, New York, NY, USA*

MARINA VENERO GALANTERNIK • *Division of Developmental Biology, Eunice Kennedy Shriver National Institute of Child Health and Human Development, NIH, Bethesda, MD, USA*

LYNNE EL GHORAYEB • *The Institute for Biomedical Sciences, The George Washington University, Washington, DC, USA*

EVE-ELLEN GOVEK • *Laboratory of Developmental Neurobiology, The Rockefeller University, New York, NY, USA*

GIANNA GRAZIANO • *Department of Ophthalmology, Johns Hopkins University School of Medicine, Baltimore, MD, USA*

ABIGAIL HAFFEY • *The Institute for Biomedical Sciences, The George Washington University, Washington, DC, USA; Center for Cancer and Immunology Research, Children's National Hospital, Washington, DC, USA*

ROWLAND H. HAN • *Department of Neurological Surgery, Washington University School of Medicine, St. Louis, MO, USA*

MARY E. HATTEN • *Laboratory of Developmental Neurobiology, The Rockefeller University, New York, NY, USA*

DALIA HAYDAR • *The Institute for Biomedical Sciences, The George Washington University, Washington, DC, USA*

TONG HU • *Division of Oncology, Department of Medicine, Washington University School of Medicine, St. Louis, MO, USA*

XIAO CHRIS HUANG • *Department of Pediatrics, Division of Experimental Hematology and Cancer Biology, Cincinnati Children's Hospital Medical Center, Cincinnati, OH, USA*

ANNA JERMAKOWIZ • *Laboratory of Developmental Neurobiology, The Rockefeller University, New York, NY, USA*

DAISUKE KAWAUCHI • *Department of Neuro-oncology, Institute of Brain Science, Graduate School of Medical Sciences, Nagoya City University, Aichi, Japan*

ASHWANI KESARWANI • *Division of Oncology, Department of Medicine, Washington University School of Medicine, St. Louis, MO, USA*

KHATEREH KHORSANDI • *The Institute for Biomedical Sciences, The George Washington University, Washington, DC, USA*

ALBERT H. KIM • *Departments of Neurological Surgery, Genetics, Neurology, Developmental Biology, Washington University School of Medicine, The Brain Tumor Center, Siteman Cancer Center, St. Louis, MO, USA*

HENG LI • *Department of Radiation Oncology and Molecular Radiation Sciences, Johns Hopkins University School of Medicine, Baltimore, MD, USA; 2E316 Sibley Memorial Hospital, Washington, DC, USA*

YUPING DEREK LI • *Division of Oncology, Department of Medicine, Department of Neurosurgery, Washington University School of Medicine, St. Louis, MO, USA*

CHARLES L. LIMOLI • *Department of Radiation Oncology, University of California, Irvine School of Medicine, Irvine, CA, USA*

Q. RICHARD LU • *Department of Pediatrics, Division of Experimental Hematology and Cancer Biology, Cincinnati Children's Hospital Medical Center, Cincinnati, OH, USA*

ZAILI LUO • *Division of Pediatric Hematology/Oncology, Children's Hospital of Alabama and the University of Alabama at Birmingham, Birmingham, Alabama, USA*

SPENCER M. MAINGI • *Department of Radiation Oncology, Duke University School of Medicine, Durham, NC, USA*

DIANE D. MAO • *Department of Neurological Surgery, Washington University School of Medicine, St. Louis, MO, USA*

DEVIN MILES • *Department of Radiation Oncology and Molecular Radiation Sciences, Johns Hopkins University School of Medicine, Baltimore, MD, USA*

JEFF S. MUMM • *Department of Ophthalmology, Johns Hopkins University School of Medicine, Baltimore, MD, USA*

NAJIBA MURAD • *Center for Cancer and Immunology Research, Children's Research Institute, Children's National Medical Center, Washington, DC, USA*

KATRINA O'HALLORAN • *Children's Hospital Los Angeles, Los Angeles, CA, USA*

DEJERIANNE OSTROW • *Children's Hospital Los Angeles, Los Angeles, CA, USA*

YANXIN PEI • *Center for Cancer and Immunology Research, Children's Research Institute, Children's National Medical Center, Washington, DC, USA*

ERIC H. RAABE • *Division of Pediatric Oncology, Sidney Kimmel Comprehensive Cancer Center, Johns Hopkins School of Medicine, Baltimore, MD, USA*

ZACHARY J. REITMAN • *Department of Radiation Oncology, Duke University School of Medicine, Durham, NC, USA*

DAVID RUBLE • *Children's Hospital Los Angeles, Los Angeles, CA, USA*

ELTON VANNOY • *The Institute for Biomedical Sciences, The George Washington University, Washington, DC, USA; Center for Cancer and Immunology Research, Children's National Hospital, Washington, DC, USA*

MARIE-CATHERINE VOZENIN • *Secteur Radio-Oncologie et Radiobiologie, Hôpitaux Universitaires de Genève, LiRR—Laboratory of Innovation in Radiobiology Applied to Radiotherapy, Faculty of Medicine, University of Geneva, Geneva, Switzerland*
KAIXUAN WANG • *Division of Pediatric Oncology, Sidney Kimmel Comprehensive Cancer Center, Johns Hopkins School of Medicine, Baltimore, MD, USA*
WANCHEN WANG • *Department of Neuro-oncology, Institute of Brain Science, Graduate School of Medical Sciences, Nagoya City University, Aichi, Japan; Graduate School of Medical and Dental Sciences, Institute of Science Tokyo, Tokyo, Japan*
WANCHEN WANG • *Department of Neuro-oncology, Institute of Brain Science, Graduate School of Medical Sciences, Nagoya City University, Aichi, Japan; Graduate School of Medical and Dental Sciences, Institute of Science Tokyo, Tokyo, Japan*
ZELUN WANG • *Department of Neurological Surgery, Washington University School of Medicine, St. Louis, MO, USA*
BRANT M. WEINSTEIN • *Division of Developmental Biology, Eunice Kennedy Shriver National Institute of Child Health and Human Development, NIH, Bethesda, MD, USA*
TIMOTHY R. WOODIWISS • *Department of Neurological Surgery, University of Iowa Carver College of Medicine, Iowa City, IA, USA*
HUANHUAN XIAO • *Department of Ophthalmology, Johns Hopkins University School of Medicine, Baltimore, MD, USA*
DAZHUAN ERIC XIN • *Department of Pediatrics, Division of Experimental Hematology and Cancer Biology, Cincinnati Children's Hospital Medical Center, Cincinnati, OH, USA*
DONG XU • *Children's Hospital Los Angeles, Los Angeles, CA, USA*
ZHENHUA XU • *Center for Cancer and Immunology Research, Children's Research Institute, Children's National Medical Center, Washington, DC, USA*
SERGE YAACOUB • *The Institute for Biomedical Sciences, The George Washington University, Washington, DC, USA*
VENKATA YELLAPANTULA • *Children's Hospital Los Angeles, Los Angeles, CA, USA*
LINGSHU YIN • *Department of Radiation Oncology and Molecular Radiation Sciences, Johns Hopkins University School of Medicine, Baltimore, MD, USA*
YURAN YU • *Department of Biochemistry and Molecular Biology, Bloomberg School of Public Health, Johns Hopkins University, Baltimore, MD, USA*
YUXIA ZHAN • *Children's Hospital Los Angeles, Los Angeles, CA, USA*

Chapter 1

Single-Cell Multi-omics Analyses of Medulloblastoma

Dazhuan Eric Xin, Zaili Luo, Xiao Chris Huang, and Q. Richard Lu

Abstract

Medulloblastoma is the most common malignant pediatric brain tumor. These tumors have considerable intratumoral and intertumoral heterogeneity. Recent advancements in single-cell multi-omics technologies have provided unprecedented insights into the cellular and molecular landscapes of medulloblastoma, enabling the identification of novel therapeutic targets and improving our understanding of tumorigenesis and disease progression. This chapter outlines essential protocols for multi-omics analyses of medulloblastoma. Utilizing a dataset from a recent multi-omics study on human medulloblastoma samples, we describe the framework for performing single-cell transcriptomics, genomics, and chromosomal interaction analysis and data-driven integration to identify the molecular networks underlying disease pathogenesis. This protocol highlights the profound impact of multi-omics technologies on our understanding of various aspects of pediatric brain tumors, including tumor lineage evolution, activated molecular pathways, cancer genetics and epigenetics, and the identification of potential therapeutic targets.

Key words Medulloblastoma, Single-cell multi-omics, Tumor heterogeneity, Data integration, 3D genome interactions, Therapeutic targets

1 Introduction

Acquiring multi-omics data has become a standard practice in life sciences, as these datasets provide detailed molecular insights into tissue function in health and disease [1, 2]. Multi-omics analyses have enhanced our understanding of brain tumor heterogeneity, evolution, and treatment options [3, 4]. Computational analyses that integrate multi-omics data have enabled the identification of key molecular features in tumors that can be mapped to knowledge bases with annotations on ontologies, pathways, and molecular interactions [5, 6]. Integrating multiple omics profiles offers insights into tumor cell states, heterogeneity, and regulatory mechanisms.

In this protocol, we provide the methods for single-cell RNA sequencing (scRNA-seq) and single-nuclei assay for transposase-accessible chromatin sequencing (snATAC-seq) analyses of fresh

Michael Goldstein (ed.), *Translational Research Methods for Pediatric Brain Tumors*, Neuromethods, vol. 226,
https://doi.org/10.1007/978-1-0716-4754-7_1,

and frozen tumor samples and describe a series of web-based tools designed to perform a wide range of omics data analysis tasks through an accessible web interface. These methods are useful for studying cancers such as medulloblastoma (MB), which is broadly comprised of Sonic Hedgehog (SHH), Group-3 (G3), Group-4 (G4), and WNT subgroups [7, 8]. Using a recently published multi-omics dataset on human MB samples [9, 10], we first illustrate the analysis of transcriptomics data. Next, we explore knowledge-driven integration approaches that leverage molecular features identified in the initial analyses. This chapter focuses mainly on the application of scRNA-seq, snATAC-seq, and Hi-C in the characterization of MB, highlighting key findings. The methodology provides a conceptual framework along with step-by-step instructions on utilizing this suite of web-based tools for comprehensive multi-omics analyses.

2 Materials

2.1 Reagents

1. Fresh and frozen MB tissue samples.
2. Trypan blue stain.
3. Chromium Single Cell 3′ Gel Bead Kit (10x Genomics).
4. Chromium Single Cell 3′ Library Kit (10x Genomics).
5. Chromium i7 Multiplex Kit (10x Genomics).
6. Chromium Single Cell ATAC Library & Gel Bead Kit (10x Genomics).
7. High Sensitivity DNA Kit (Agilent).
8. Arima Hi-C Kit (Arima Genomics).

2.2 Equipment

1. 10× Genomics Chromium Controller.
2. Agilent 2100 BioAnalyzer system.
3. Bio-Rad C1000 Touch Thermal Cycler.
4. Illumina NovaSeq 6000.

2.3 Computing System

1. CentOS Linux-based HPC cluster.
2. Windows PC 32G RAM.

2.4 Software

1. R version 4.0.3 and 4.0.5, RStudio.
2. Cellranger, Seurat, Monocle 3, STREAM, VECTOR, Slingshot, CIBERSORTx, inferCNV, ArchR, Juicer (version 1.5), HiCCUPS, NeoLoopFinder.

3 Methods

3.1 General Workflow for scRNA-seq and snATAC-seq of MB Samples

scRNA-seq enables high-throughput transcriptional profiling through analysis of gene expression at the individual cell level, offering insights into cellular heterogeneity within complex samples such as MB tissues. The process begins with isolating cells from fresh tissues and/or cryopreserved cells and quality assessment using the 10x Genomics Chromium Controller. After cell lysis, mRNA molecules are captured, reverse-transcribed into cDNA, and tagged with unique cell barcodes and molecular identifiers (UMIs). The resulting cDNA is amplified and linked to Illumina adapters for sequencing. The scRNA-seq data are subjected to quality control and doublet removal as described [11] followed by cell population clustering to enable comprehensive single-cell transcriptomic analysis. An overview of the protocol workflow is illustrated in Fig. 1.

scRNA-seq and snATAC-seq datasets of matched tissues from tumors from 26 MB patients were analyzed to define the cellular identities of cerebellar MBs. By comparing human fetal cerebellar cell type profiles, we identified a newly identified populations similar to transitional cerebellar progenitors (TCPs) in G3, G4, and SHH MBs as well as tumor subtype-specific cell clusters (Fig. 2a).

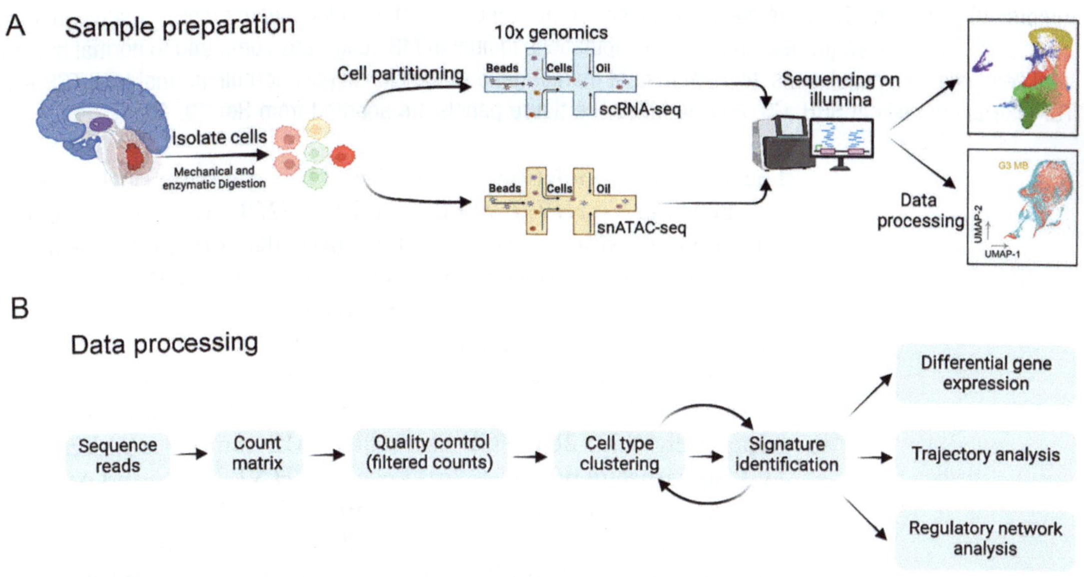

Fig. 1 Singe-cell transcriptomic data generation and analysis. (**a**). Schematic depicting sample processing for single-cell transcriptomics. Individual cells are dissociated from tissues and encapsulated in oil-based droplets for RNA amplification and library construction by 10× Genomics system. After data processing, gene expression in individual cells is analyzed by clustering. (**b**). The scRNA-seq data are subjected to quality control and removal of doublet and low-quality cell data, followed by cell population clustering to enable comprehensive single-cell transcriptomic analysis

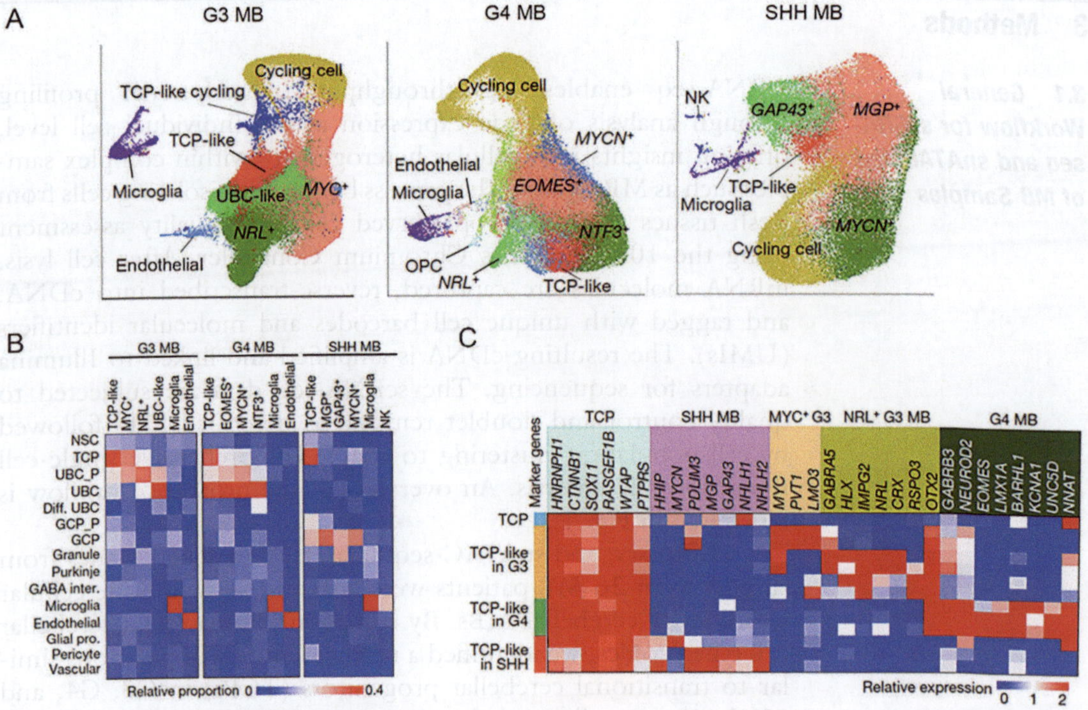

Fig. 2. Singe-cell transcriptomics reveals intermediate TCP-like progenitors in MBs. (**a**), UMAP visualizations of transcriptomes of G3 ($n = 9$ tumors; 90,672 cells), G4 ($n = 13$ tumors; 61,254 cells), and SHH MB ($n = 4$ tumors; 45,731 cells). The cells are color-coded for indicated populations identified by representative marker genes. NK, nature killer. (**b**) Heat map of cell populations in human MB subgroups compared to normal human fetal cerebellar cell populations. (**c**) Heat map of marker gene expression in pseudo-bulk normal fetal TCP and TCP-like tumor cells in SHH, G3, and G4 MBs. The figure panels are adapted from Ref. [9, 10]

The TCP cell population, characterized by expression of the signature genes *HNRNPH1*, *SOX11*, and *CTNNB1*, is highly enriched within the transitional zone and the ventricular zone in the rhombic lip region at post conception week 12 [9, 10]. TCP-like cells in different MB subgroups transcriptionally mimic the normal TCP populations, whereas neoplastic cells in G3 and G4 MBs and in SHH MBs also had gene signatures similar to those of the unipolar brush cell (UBC) lineage and of granule cell precursors (GCPs), respectively (Fig. 2b). TCP-like cells in MB tumors exhibited the TCP signature genes (Fig. 2c). Levels of *HNRNPH1* and *SOX11* mRNAs are elevated in a variety of cancers [12, 13], and *CTNNB1* encodes a tumor driver in the cerebellum if normal lineage restriction is lifted [14]. TCP-like populations in G3, G4, and SHH MBs also expressed subgroup-specific signatures (Fig. 2c). Exemplary results of the resulting single-cell multi-omics of medulloblastomas are shown in Fig. 2.

10× Genomics scRNA-seq Data Acquisition

scRNA-seq was performed on the 10x Genomics platform according to the manufacturer's instructions.

(a) Determine the concentration of single cells in suspension by staining with Trypan blue (0.4%). Ensure that more than 90% of cells are viable.

(b) Load approximately 10,000–30,000 cells per sample on the Chromium Controller to generate single-cell GEMs.

(c) Perform GEM reverse-transcription, DynaBead cleanup, PCR amplification, and SPRIselect bead cleanup using the Chromium Single Cell 3′ Gel Bead kit (10× Genomics).

(d) Generate indexed single-cell libraries using the Chromium Single Cell 3′ Library kit and the Chromium i7 Multiplex kit (10× Genomics).

(e) Evaluate cDNAs and libraries for size, quality, concentration, and purity using the Agilent 2100 Bioanalyzer system using a High Sensitivity DNA Kit (Agilent Technologies).

Perform sequencing in an Illumina sequencer (e.g., NovaSeq 6000).

10× Genomics snATAC-seq Data Acquisition

snATAC-seq was performed on the 10x Genomics platform according to the manufacturer's instructions.

(a) Incubate isolated single cells at 37 °C for 1 h with 10 μl of transposition mix (per reaction 7 μl ATAC Buffer and 3 μl ATAC Enzyme; 10x Genomics).

(b) Load approximately 10,000 cells per sample on the Chromium Controller to generate single-cell GEMs.

(c) Perform reverse transcription in a C1000 Touch Thermal Cycler (Bio-Rad) programmed at 72 °C for 5 min, 98 °C for 30 s, 12 cycles of 98 °C for 10 s, 59 °C for 30 s, and 72 °C for 1 min; hold at 15 °C.

(d) Perform DynaBead cleanup, PCR amplification, and SPRIselect bead cleanup using the Chromium Single Cell 3′ Gel Bead kit (10× Genomics).

(e) Generate indexed single-cell libraries using the Chromium Single Cell 3′ Library kit and the Chromium i7 Multiplex kit (10x Genomics.

(f) Evaluate cDNAs and libraries for size, quality, concentration, and purity using the Agilent 2100 Bioanalyzer system using a High Sensitivity DNA Kit (Agilent Technologies).

(g) Perform sequencing on NovaSeq 6000 at a depth of approximately 400,000,000 reads per sample.

scRNA-seq Data Processing and Quality Filtering

(a) Demultiplex by converting raw BCL sequencing files into FASTQ files using Cellranger (https://github.com/10XGenomics/cellranger).

(b) Align the FASTQ files to the human reference genome (GRCh38) using Cellranger to generate gene-barcode count matrices.

(c) Aggregate multiple samples using Cellranger aggr. This normalizes for sequencing depth across samples and produces an aggregated gene-barcode matrix.

(d) Import the gene-barcode matrix into Seurat (https://satijalab.org/seurat/), an R package for scRNA-seq analysis.

(e) Filter out low-quality cells by removing cells that express fewer than 200 genes and more than 5000 genes and by excluding cells with high mitochondrial gene content (more than 5 standard deviations above the median).

(f) Remove doublets using the DoubletFinder package [10] or DropletQC [15].

(g) Remove genes expressed in fewer than three cells.

(h) Normalize gene expression data using Seurat's LogNormalize method, which adjusts for differences in sequencing depth and then log-transforms the data.

(i) Scale the data by adjusting gene expression values for each cell, ensuring comparable data across all genes.

(j) Perform dimensionality reduction: Perform principal component analysis (PCA) to reduce data complexity, a necessary step for clustering and visualization.

Clustering Analysis, Visualization, and Annotation of scRNA-seq Data

(a) Default parameters of Seurat V4 were used if not otherwise stated. Using the Seurat pipeline, calculate the average expression and dispersion for each gene, and then group the genes based on their average expression. Briefly, compute the z-score for dispersion within each group and apply a z-score cutoff of 0.5, and filter the genes by setting an average expression cutoff (e.g., within the range of 0.0125 to 3.0).

(b) Perform linear dimensionality reduction using PCA. Select the statistically significant principal components by applying the elbow method and the jackstraw method from the Seurat package.

(c) Cluster cells by applying the shared nearest neighbor modularity optimization-based clustering algorithm in Seurat.

(d) Visualize the identified clusters using t-distributed stochastic neighbor embedding (t-SNE) and uniform manifold approximation and projection (UMAP); both are part of the Seurat package.

(e) Assign cell-type identities to clusters through manual annotation using known marker genes for expected cell types.

(f) Identify differentially expressed genes. Ensure that genes to be analyzed are expressed in at least 25% of cells in one of the two compared populations. Set a minimum natural log scale expression difference of 0.2. Use Seurat's FindAllMarkers function to identify differentially expressed genes. Perform a one-tailed Wilcoxon rank sum test to calculate statistical significance. Adjust p-values for multiple testing using the Bonferroni correction.

(g) Perform manual annotation iteratively, validating the proposed cell clusters using known markers. If gene signatures indicate additional diversity within clusters, investigate these clusters for potential subpopulations.

3.2 Pseudotime Cell-State Trajectory Analysis Based on scRNA-seq Data

To investigate cell lineage trajectories in human fetal cerebellar neural cell types, we employed Monocle analysis [16]. Based on the stemness score, we identified the neural stem cell (NSC) population as the starting point. We observed a trajectory through TCPs to the three main neuronal lineage branches: GCPs, UBCs, and Purkinje cells (Fig. 3a). To define expression dynamics along the trajectory, we used Slingshot [17] and found successional gene expression from NSCs to TCPs (Fig. 3b). STREAM [18] confirmed this hierarchy (Fig. 3c). Unsupervised VECTOR trajectory analysis [19] predicted that TCP-like cells were in a less differentiated state (Fig. 3d). The procedure used for this analysis was as follows:

(a) Perform cell clustering using UMAP.

(b) Use an unsupervised method such as Monocle 3 [20, 21], STREAM [18], or VECTOR [19] to identify starting cells and to infer cell-state transition directions. These methods do the following:

- Normalize the barcodes of selected clusters (this can be performed using either Monocle dPFeature or Seurat).
- Remove genes with low expression.
- Perform PCA on the remaining genes to select significant principal components.
- Use a generalized linear model to identify differentially expressed genes.
- Select the top 1000 genes per cluster or use the default parameters within each software.

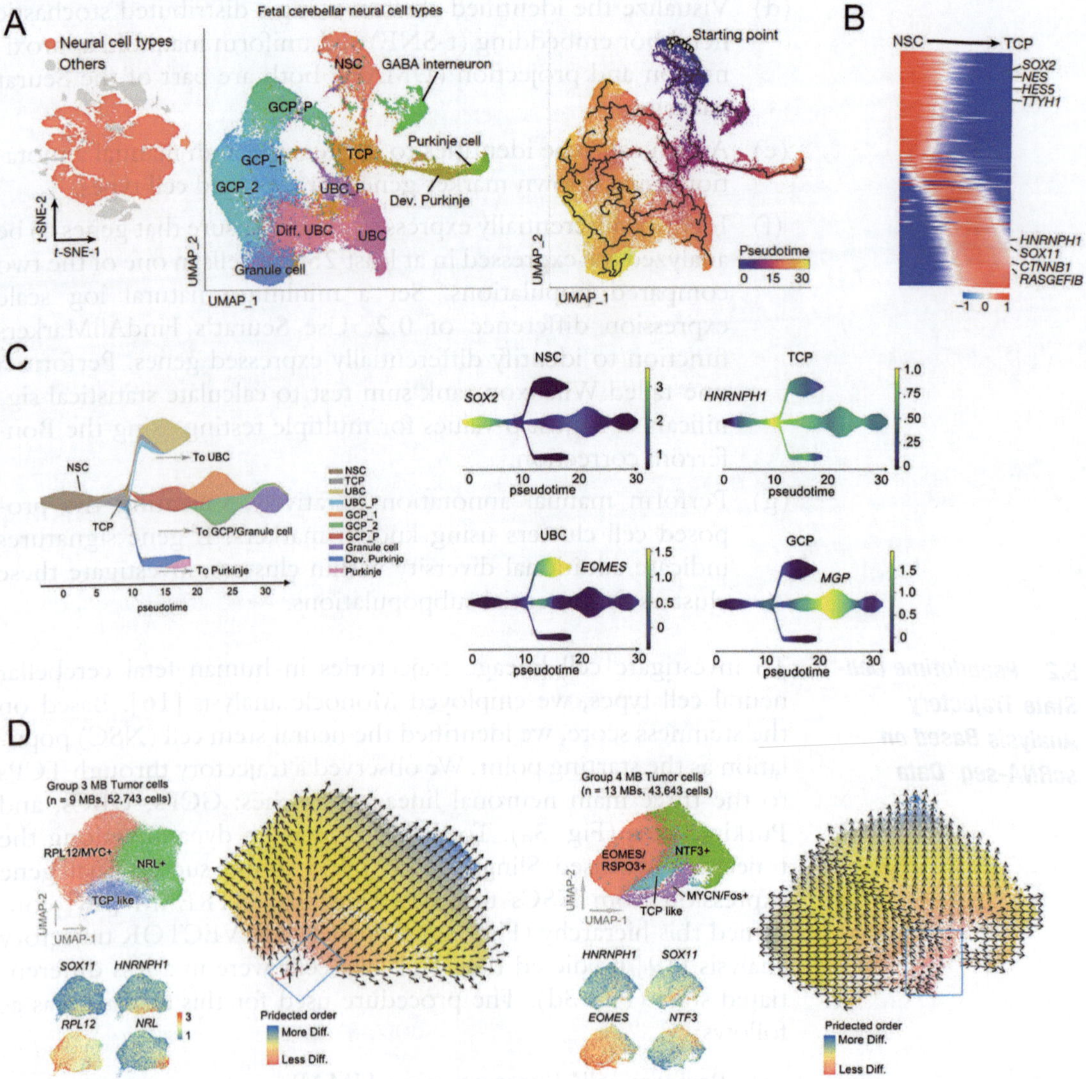

Fig. 3. Single-cell transcriptomic profiling of early developing human fetal cerebella. (**a**) Left, neural cell types (red) in the *t*-SNE map of fetal cerebella. Middle, UMAP visualization of neural cell types from fetal cerebella. Right, Prediction of neural cell type fate transitions using Monocle 3. Dev. Purkinje cell, developing Purkinje cell. (**b**), Heat map with spline curves fitted to differentially expressed genes. Pseudo-temporal trajectories were grouped by hierarchical clustering. (**c**) Left, STREAM plot visualization of cell density along the indicated cell lineage trajectories. Right, STREAM visualization of inferred branching points, trajectories, and expression of key genes in NSCs, TCPs, UBCs, and GCPs. (**d**), Population, marker gene expression, and trajectory analysis of human G3 and G4 MBs using Vector based on the UMAP images. The figure panels are adapted from Ref. [9, 10]

- Principal Graph Fitting: Fit these genes into a principal graph within each partition using the learn_graph function in the Seurat package.

(c) Slingshot Cellular Trajectory Analysis [17]:

Filter and normalize the input matrix and annotate cell types and provide these annotations as labels for Slingshot, which is part of the Seurat R package.

3.3 Tumor Deconvolution Analysis

Tumor deconvolution analysis aims to computationally infer the proportions and identities of different cell types present within bulk tumor samples. Since tumor tissues consist of a heterogeneous mixture of cancerous, stromal, immune, and other cell types, traditional bulk transcriptomics or genomic profiling often masks the contributions of individual cell populations. We recently used CIBERSORTx analysis [22] to show that TCP-like populations are present in higher abundances in G3 MBs than in G4 and SHH MBs (Fig. 4a). The procedure used for this analysis was as follows:

(a) Prepare the Input Data:

Tumor Data: Collect bulk RNA-seq data or clusters of scRNA-seq data from tumor samples.

Signature Matrix: Prepare the human fetal cerebellum cell cluster expression matrix.

(b) Gene Filtering: Remove genes related to cell cycle (~1400 genes), ribosome biogenesis (~300 genes), mitochondrial genes, and apoptosis-related genes (~100 genes) to reduce bias.

(c) Upload the cluster expression matrix to CIBERSORTx (https://cibersortx.stanford.edu/runcibersortx.php).

Input Mixture: Upload the transcriptomes of the tumor data (bulk RNA-seq or scRNA-seq clusters) to the CIBERSORTx platform.

Signature Matrix: Upload the prepared human fetal cerebellum cluster expression matrix (maximum of each gene within each group average expression >0.1) [9] as the signature matrix.

(d) Configure Analysis Settings:

Normalization: Disable quantile normalization to avoid standardizing the data.

Permutations: Set the number of permutations for significance analysis between 100 and 500.

(e) Submit the job to perform the deconvolution analysis.

(f) Download and review the results, which will include the estimated cell type proportions in the tumor samples.

(g) Analyze the deconvolution output to understand the cell type composition of the tumor samples based on the human fetal cerebellum clusters.

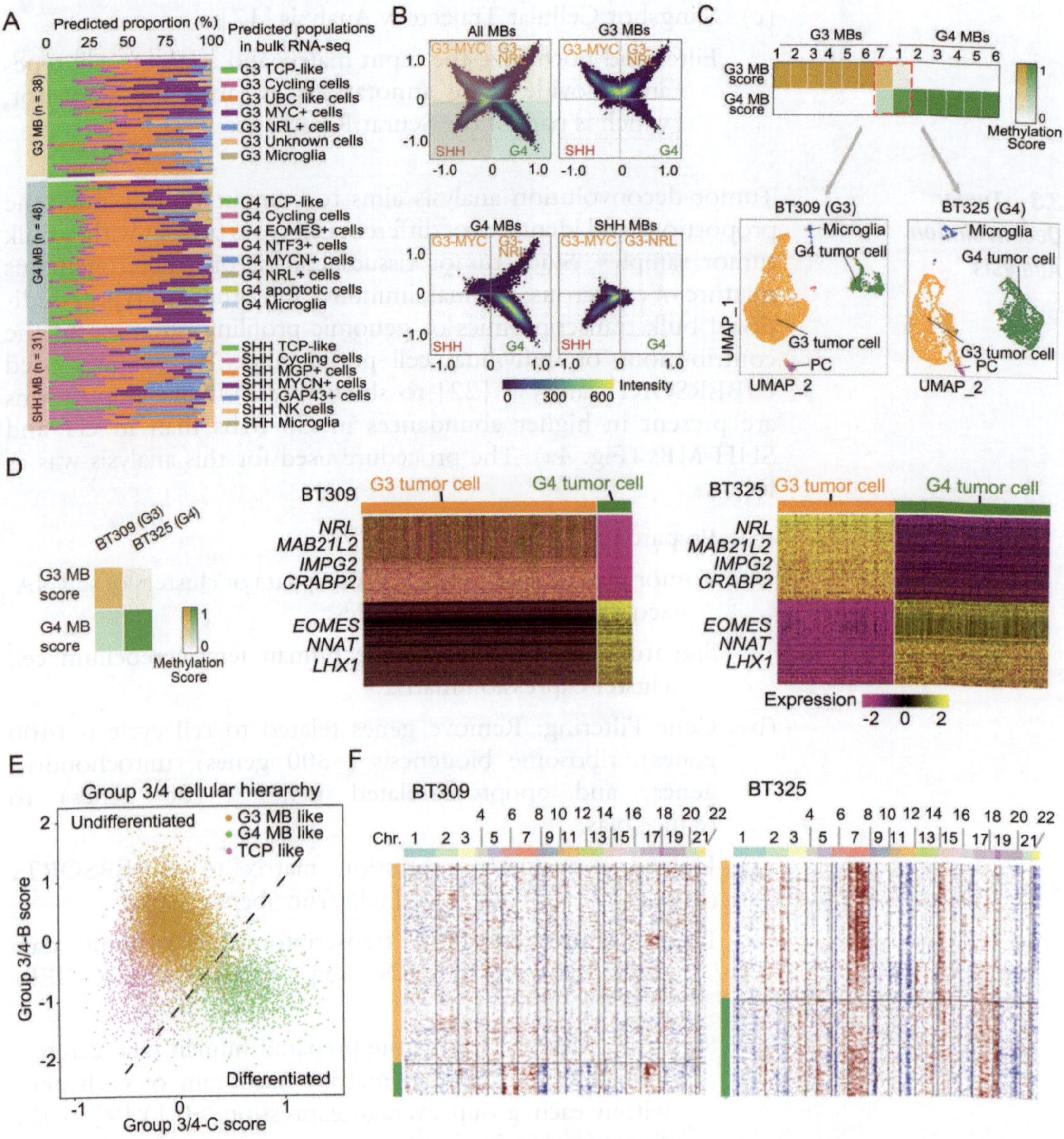

Fig. 4. Cell populations and cell-state plots of tumor cells in MB subgroups. (**a**), Fractions of cells in human MB subgroup CBTTC cohorts predicted using CIBERSORTx deconvolution; SHH, G3 and G4 MB tumor cell types were used as references. (**b**) Cell-state plots of tumor subpopulations in indicated MB subgroups. The positions of cell states (dots) indicate relative meta-module scores, and color scales represent the count. (**c**) Top, DNA methylation-based subgroup prediction scores for G3 and G4 MBs. Samples G3-7 (BT-309) and G4-1 (BT-325) had a mixture of G3/G4 scores. Bottom, UMAP distributions of G3 and G4 tumor cells from BT-309 and BT-325. PC, photoreceptor cells. (**d**) Heatmap of the marker gene expression in BT-309 and BT-325 tumor cells showing mixed G3 and G4 cell populations. (**e**), Left, DNA methylation-based subgroup prediction for G3 (BT-309) and G4 (BT-325). Right, Scatter plots of mixed G3 and G4 MB tumor cells based on Group 3/4 metaprogram expression scores. (**f**), Heatmap of CNV signals normalized against the "Microglia" cluster shows CNV changes by chromosome (columns) for individual cells (rows) based on scRNA-seq data. The figure panels are adapted from Ref. [9, 10]

3.4 Two-Dimensional Tumor Cell States, Tumor Subgroup Scoring, and Copy Number Variation Analysis

G3 and G4 MBs share similar signature genes based on bulk transcriptome and DNA methylation profiles [23, 24]. Identification of differences in tumor cell states and cellular populations in the tumor microenvironments between G3 and G4 MBs is essential for understanding intratumoral and intertumoral heterogeneity, which should reveal why G3 and G4 MBs, which have with similar histological profiles, have very different clinical outcomes. Cell-state analyses reveal the extent of tumor cell plasticity, which is the dynamic ability of cancer cells to transition between states (e.g., stem-like, quiescent, or proliferative), which contributes to metastasis, therapy resistance, and recurrence. Cell-state analysis can guide tumor classification and can inform strategies for immune modulation and chemotherapy development, ultimately leading to improved patient outcomes.

Cell-state plots [25] revealed specific enrichment of tumor cell states associated with MB subgroups. These analyses indicated that G3 and G4 tumors have cells with either mixed or intermediate plasticity states. In contrast to the relatively restricted tumor cell state in SHH MB, we detected MYC^+ and NRL^+ cell states as well as a state with a G4 MB-like signature in G3 MBs (Fig. 4b). The G4 MB tumors were enriched in G4 MB-specific states and a cell state with an NRL^+ G3 signature (Fig. 4b). To evaluate the hypothesis that intermediate cell populations are shared between G3 and G4 tumors, we analyzed samples from two patients (BT-309 and BT-325) that showed a mixture of G3 and G4 signatures based on a DNA methylation profiling (Fig. 4c). Unbiased single-cell clustering of transcriptomes and cell-state assignment analysis of Group 3/4 MB metaprograms revealed distinct G3 and G4 tumor cell populations; we did not detect intermediate states expressing both G3 and G4 signatures in significant frequencies (Fig. 4d, e). Single-cell copy number variation (CNV) analysis confirmed that G3- and G4-like populations with distinct patterns were present within individual tumors (Fig. 4f). These data suggest that a set of G3 and G4 MB tumors harbor a mixture of G3- and G4-specific cell populations rather than a G3-to-G4 cell-state continuum. These analyses that led to these conclusions were performed as follows:

Identification of Tumor Cell States and Tumor Subgroup Scoring

(a) Prepare the tumor cell cluster dataset for analysis by separation into the following categories: G3-MYC, G3-NRL, G4, and SHH.

(b) Compute subtype expression scores:

- Define Nearest Neighbors: For each tumor cell in the single-cell transcriptomics dataset, determine its 100 nearest neighbors based on Euclidean distance.

- Calculate Subtype Fraction: For each cell, compute the fraction of its 100 nearest neighbors that belong to each of the predefined clusters (e.g., G3-MYC, G3-NRL, G4-, and SHH).
- Record these fractions for subsequent visualization.
- Visualization of Enrichment: Create a two-dimensional representation of the tumor cells using t-SNE or PCA.
- Color Coding: Use color coding to display the fractions calculated in the previous step. Colors representing varying levels of enrichment for each cluster subset.

(c) Analyze cell states and cell distribution in G3 or G4 MB.

- Identification of signature genes for tumor subgroups from bulk tissues: The top 30 genes per group (e.g., G3 and G4 MB) from each previously defined subtype-specific meta-program [26] based on DNA methylation data from bulk Group 3/4 MBs are utilized to calculate subtype expression scores for each tumor cell in the datasets.
- Align the single-cell transcriptomic data with the signature genes (e.g., 30 genes per G3 or G4 MB) associated with Group 3/4 MB metaprograms: The relative expression of the signature genes associated with Group 3/4 MB meta-programs serves as a matrix reference for defining cell types in bulk Group 3/4 MBs. This matrix is used to align the single-cell data, enabling the assignment of each cell to distinct or overlapping cell states.
- Visualize the scores of each tumor cell by plotting the distribution and enrichment of tumor cell subtypes.

Inferred CNV Analysis From scRNA-seq

(a) Obtain scRNA-seq data from the samples of interest, and ensure that the data are correctly formatted for use with inferCNV [27], adhering to the required input structure and file formats.

(b) Preprocess data:

- Prepare the expression profiles for use with inferCNV, ensuring data is organized by chromosomal intervals.
- Use moving averages to smooth expression profiles across chromosomal intervals.

(c) Run inferCNV to infer large-scale chromosomal variations in each single cell. CNV signals, normalized against the CNV "Microglia" cluster, reveal copy number alterations.

Integration of CNV-Based Classification with Transcriptomic Analysis

(a) Perform CNV-based classification.

- Classify cells as malignant if they display large-scale chromosomal variations.
- Classify cells as nonmalignant if they have no CNVs are considered non-malignant.

(b) Perform transcriptome-based clustering, identifying clusters corresponding to malignant and nonmalignant cells based on CNV profiles.

(c) Integrate CNV and transcriptomic data:

- Integrate CNV classification with transcriptome-based clustering of cells and expression of nonmalignant marker genes to identify malignant (with CNV) and nonmalignant cells.
- Classify malignant and nonmalignant categories based on combined CNV and marker gene expression data.

Malignant Cells: exhibit CNV profiles without high expression of nonmalignant marker genes.

Nonmalignant Cells: Selected as normal cells and assumed to be a reference without CNV profiles.

3.5 Identification of Networks That Drive TCP Transformation

To decipher how dynamic accessibility at *cis*-regulatory elements (CRE) relates to the gene regulatory programs in TCP-like cells from aggressive MBs, we performed snATAC-seq of G3 and G4 MB samples for which matched scRNA-seq data were available (Fig. 5a–c). The correlations between the accessibility of promoter and gene body elements with target gene expression were calculated using ArchR [28]. We found that the positively correlated peak-to-gene pairs were mostly cell cluster-specific (Fig. 5d, e). To determine the temporal relationship between chromatin accessibility and gene expression, we ordered the peak-to-gene pairs based on their CRE accessibility as a function of pseudotime. In G3 tumors, the TCP-like cluster preceded the *MYC*$^+$ cell cluster, which was followed by the *NRL*$^+$ cell cluster (Fig. 5f). Transcription factor motif analysis indicated an enrichment of accessibility in binding motifs for SOX11 and TWIST1 in the TCP-like cells, for TCF3 and MYC motifs in the *MYC*$^+$ G3 cells, and for NR2F1 and PAX5 motifs in the *NRL*$^+$ G3 cells (Fig. 5f). Gene ontology analysis of G3 MB clusters showed an enrichment of epithelial development, epithelial-to-mesenchymal transition, and TGFb/BMP signaling in the TCP-like population (Fig. 5f). In G4 MBs, the accessibility in TCP-like cells emerged prior to that of the *KCNA1*$^+$ and *EOMES*$^+$ cell clusters. An enrichment of epithelial cell development, neural progenitor cells, and UBC signatures was observed in TCP-like cells in G4 tumors (Fig. 5f), whereas the *EOMES*$^+$ G4 subpopulation was enriched in neuronal development

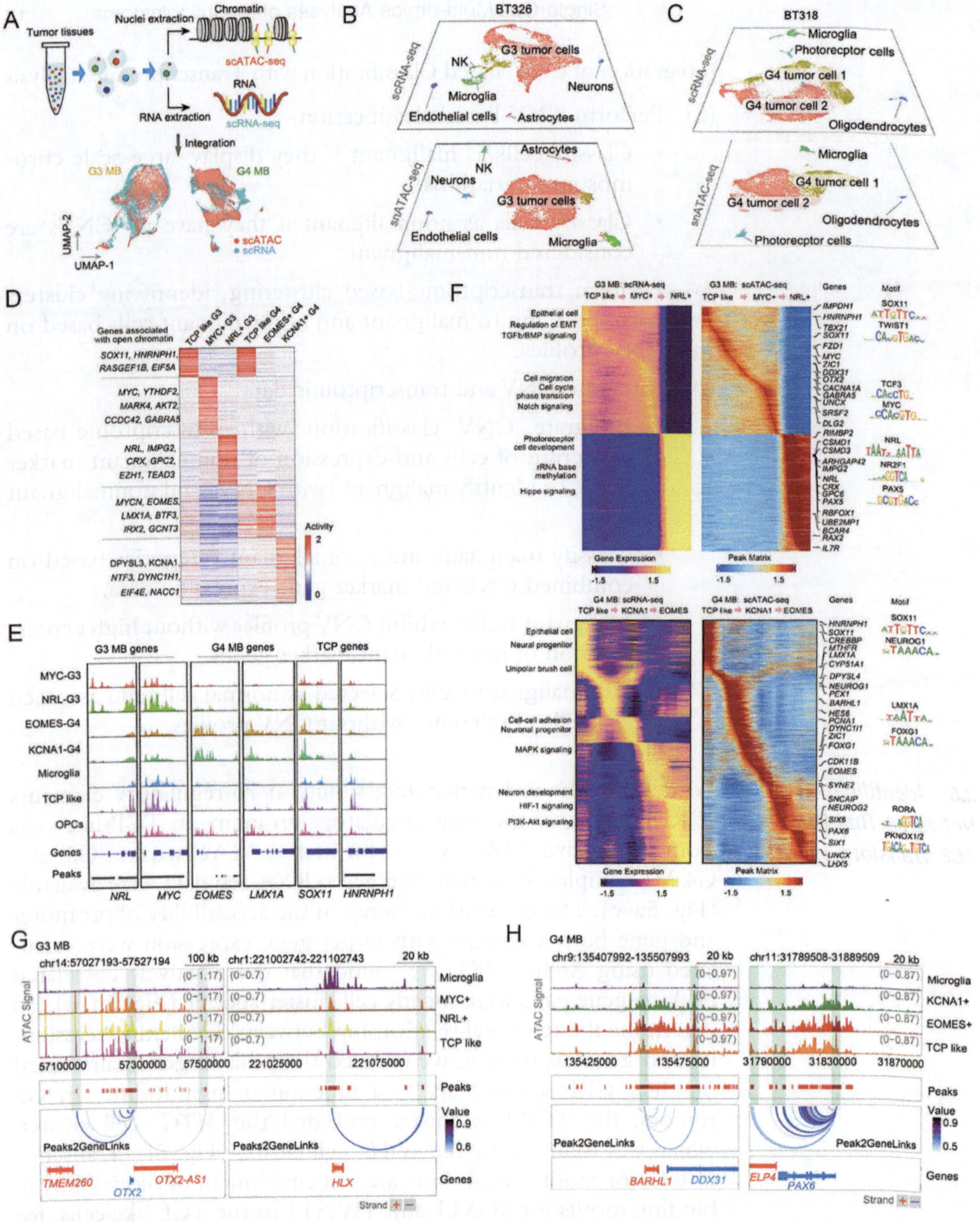

Fig. 5. Identification of regulatory networks that drive TCP transformation. **(a)** Workflow for integrating scRNA-seq and snATAC-seq data analysis using ArchR. **(b, c)**, Comparation of the cell types in the G3 tumor tissues from **(b)** patient sample BT326 and **(c)** patient sample BT3118 detected by scRNA-seq (upper) and by snATAC-seq (lower) using Seurat and Signac analyses. **(d)** Heatmap showing correlation of accessibility of promoter regions with target gene expression for tumor cell types. **(e)** Pseudo-bulk ATAC-seq tracks at the indicated gene loci in G3 and G4 MB cells. **(f)** Heatmaps of scaled accessible peak-to-gene links identified along the integrated pseudotime trajectory from TCP-like to G3 MB (upper) and TCP-like to G4 MB (lower) subclusters. Enriched transcription factor motifs are also shown. **(g, h)**, Genome accessibility track visualization of marker genes with peak co-accessibility in **(g)** TCP-G3-like cells in G3 MB (OTX2 and HLX) and **(h)** TCP-like cells in G4 MB (BARHL1 and PAX6). The figure panels are adapted from Ref. [9, 10]

and in HIF-1 and PI3K signaling (Fig. 5f). In the *EOMES*⁺ G4 subpopulations, FOXG1 and LMX1A motifs were enriched (Fig. 5f).

To identify positive transcriptional regulators that control gene expression in TCP-like cell populations, we integrated snATAC-seq and scRNA-seq data to identify transcription factors with gene expression scores positively correlated with changes in accessibility of corresponding motifs [28]. *Cis*-regulatory sites co-accessible with the promoters of potential drivers for MBs, including *OTX2* and *HLX*, were detected in TCP-like cells in G3 tumors, whereas *BARHL1* and *PAX6* were enriched in TCP-like cells in G4 tumors (Fig. 5g, h). Co-accessibility of *cis*-regulatory elements with target gene loci might contribute to expression of MB subtype-specific drivers and their oncogenic programs. These analyses were performed as follows:

Identify Cis-Regulatory Elements and Regulatory Networks

(A) Filtering Cells by TSS Enrichment Value and Number of Unique Fragments of the snATAC-seq

(a) Acquire transcriptional start site (TSS) positions using the Bioconductor package from TxDb.Hsapiens.UCSC.hg38.knownGene.

(b) Correct for Tn5 transposon insertions by a constant offset to the read start ("+" stranded +4 bp, "–" stranded −5 bp) to adjust for any biases in insertion data [29].

(c) Define aggregation range. For our analysis, we identified insertions ±2000 base pairs (bp) relative to each unique TSS genome-wide. The analysis should ensure to correct for TSS strand orientation.

(d) Calculate TSS enrichment:

- Create a smoothed profile of the aggregated Tn5-corrected insertions around each TSS.
- Measure TSS enrichment as the maximum value of the smoothed profile.

(e) Filter snATAC-seq profiles by retaining only those profiles with at least 1000 unique fragments and TSS enrichment values of ≥0.5.

(f) Exclude snATAC-seq profiles with more than 100,000 unique nuclear fragments to reduce the potential contribution of doublets.

(g) Ensure that all profiles meet the defined criteria and are free from potential contamination or doublets.

(h) Save the number of profiles before and after filtering and any observations about the data quality.

(B) Gene Regulatory Network and Motif Enrichment Analysis

(a) Identify Gene Regulatory Modules: Use the Single-Cell Regulatory Network Inference and Clustering (SCENIC) package [30] to identify gene regulatory modules with *cis*-regulatory binding motifs in scRNA-seq and snATAC-seq datasets.

(b) Create Common Peak Set: Load the peak coordinates for merged scRNA-seq and snATAC-seq datasets into SCENIC, and convert the peak coordinates to genomic ranges using the GenomicRanges::reduce function to create a common, non-redundant set of peaks for downstream analysis.

(c) Use the ArchR package for integrated analysis of scRNA-seq and snATAC-seq datasets:

- Remove low-quality cells based on default ArchR thresholds.
- Perform dimensionality reduction using ArchR's default methods (e.g., LSI for snATAC-seq).
- Visualize data in a genome browser using ArchR's built-in tools.
- Preprocess the scRNA-seq data and annotate cells according to ArchR default parameters.
- Process snATAC-seq data to assess chromatin accessibility.
- Integrate and visualize scRNA-seq and snATAC-seq data in a unified space.
- Identify regions of differential chromatin accessibility and perform motif enrichment analysis using ArchR defaults.

(d) Utilize ArchR to identify enriched transcription factors:

- Calculate gene scores based on summed chromatin accessibility, normalized to a user-defined constant (default of 10,000).
- Correlate gene scores with chromVAR transcription factor deviation z-scores in the ArchR package.

(e) Nominate marker genes and potential tumor drivers:

- Based on the gene scores and positive TF-regulators identified from ArchR, the top 30 TFs or highly expressed genes are nominated.
- Exclude noncoding genes or ribosomal proteins from the list.

These nominated genes may serve as potential drivers or marker genes in the tumor subgroup.

3.6 Hi-C Three-Dimensional Architecture Analysis for Identification of Long-Range Enhancer Hijacking in G3 MB

To determine whether SOX11/HNRNPH1-occupied enhancers correspond to distal regulatory elements of G3 MB driver genes, we performed Hi-C chromosome conformation capture in patient-derived G3 MB (MB-004) and G4 MB (UPN3550) cells. We detected unique genomic looping in each line (Fig. 6a, b). We used NeoLoopFinder [31] to reconstruct local Hi-C maps surrounding breakpoints. We identified unique structural variations and distinct sets of the interacting genomic loci involved in neo-loop formation at loop anchors in G3 and G4 MB cells. The neo-loop formation through interchromosomal translocation in G3 MB cells placed potential promoter/enhancer elements on a chromosome 11 segment close to the promoter of *PPP1R14A* on chromosome 19 (Fig. 6c). The phosphatase encoded by *PPP1R14A* has been shown to drive oncogenic RAS signaling in human cancers [32]. Hi-C analysis also indicated that the topologically associated domains in G3 MB cells harbored unique long-distance interactions with the enhancer and promoter regions of *MYC* (Fig. 6d). By integrating the long-range interacting sites with enhancer occupancy of HNRNPH1/SOX11 in G3 MB cells and NSCs, we detected chromatin interaction loops linking the distal super-enhancer clusters bound by HNRNPH1/SOX11 upstream of *CASC8* to the active promoter/enhancer elements of the *MYC* locus in the G3 MB cells but not in NSCs (Fig. 6d). This long-range interaction loop was not detected in G4 MB cells (Fig. 6d), suggesting that a unique enhancer-hijacking event in the *MYC* locus drives oncogenesis in G3 MB. Our data thus suggest that interchromosomal translocation or structural variations might activate oncogenic drivers through enhancer hijacking in G3 MBs. This analysis was performed as follows:.

(a) Sample preparation and Hi-C library construction.

- Cell cross-linking.
  - Incubate 5×10^6 MB004 (G3 MB) and UPN3550 (G4 MB) cells with 1% formaldehyde for 10 min at room temperature to crosslink chromatin.
  - Add 0.2 M glycine to quench the formaldehyde and incubate for 5 min at room temperature.
  - Centrifuge the cells at 1000× *g* for 5 min at 4 °C and discard the supernatant.
- Cell lysis and nuclei preparation.
  - Resuspend the cell pellet in cold PBS, centrifuge at 1000× g for 5 min at 4 °C, and discard the supernatant.
  - Resuspend one reaction of crosslinked cells in 20 μL of Lysis Buffer (Arima Genomics) in a tube or a well of a PCR plate, and incubate at 4 °C for 15 min.
  - Resuspend one reaction of purified crosslinked nuclei in 20 μL of water in a PCR tube.

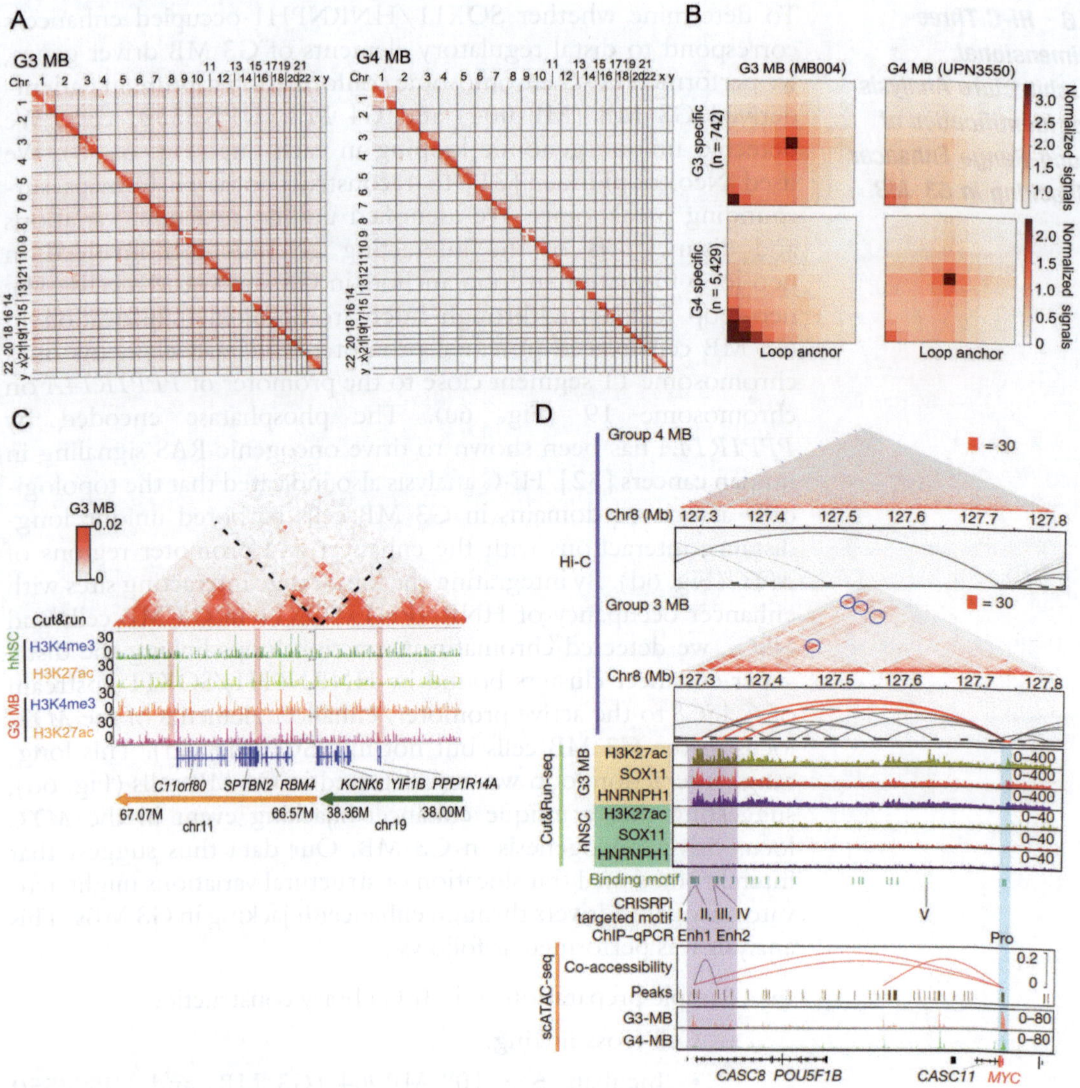

Fig. 6. 3D mapping of intrachromosomal interactions to identify long-range enhancer hijacking. (**a**) Heatmaps of Hi-C interactions across genomes of G3 (left) and G4 (right) MBs. (**b**) Aggregate peak analysis of loops in G3 and G4 MB cells. (**c**) Enhancer hijacking events identified in G3 MB-004 cells based on structural variant analysis. Inter-chromosomal translocation is observed between breakpoints chr11: 66,570,000 to 67,320,000 and chr19: 37,630,000 to 38,360,000. Genomic tracks are shown for H3K27ac/H3K4me3 occupancy in human NSCs and G3 MB cells. (**d**) Top: Hi-C map of intrachromosomal interaction in G3 MB cells (MB-004) and G4 MB cells (UPN3550). Middle: Genomic occupancies in G3 MB cells and human NSCs. Bottom: Pseudo-bulk genome co-accessibility in the MYC locus. Long-distance regulatory elements in human NSCs and G3 MB cells are highlighted. The figure panels are adapted from Ref. [9, 10]

- Add 24 μL of Conditioning Solution (Arima Genomics), and incubate at 62 °C for 10 min. Then add 20 μL of Stop Solution 2 (Arima Genomics), and incubate at 37 °C for 15 min.
- Perform in situ digestion of chromatin using the Arima Hi-C Kit according to the manufacturer's manual (Arima Genomics, Cat# A510008).

– Library preparation:
  - Follow the Arima Hi-C Kit instructions for end repair and polyA tailing.
  - Ligate the chromatin fragments using the Hi-C Kit.
  - Purify the ligation products using the kit's recommended method.
  - Amplify the Hi-C libraries using the KAPA Hyper Prep Kit (KAPA, Cat# KK2802) according to the manufacturer's protocol.

– Sequencing.
  - Quantify and normalize the libraries for sequencing.
  - Perform sequencing of the Hi-C libraries on a NovaSeq 6000 instrument (Illumina) using a 2 × 150 bp paired-end read configuration.

(b) Data processing.

– Use Juicer to process the raw sequencing reads and align them to the reference genome (hg38).

– Produce Hi-C contact matrices (.hic files) using Juicer.

– Visualize contact matrices in Juicer.

(c) Data analysis.

– Perform filtering and loop calling.
  - Use BAM files from sequencing data as input.
  - Remove low-quality reads from the BAM files.
  - Call Loops: Use Peakachu to identify loops in Hi-C data from both G3 MB and G4 MB cell lines.
  - Use diffPeakachu to compare loops between the G3 MB and G4 MB cell lines.
  - Merge subtype-specific loops using BEDTools pair-to-pair function with a negative slop of 10 kb.

– Structural Variant Identification: Run Hi-C breakfinder pipeline to detect large structural translocations, deletions, and inversions.

– Neo-Loop and Enhancer Hijacking: Neo-Loop Finder: Utilize the NeoLoopFinder pipeline to identify neo-loops or enhancer hijacking events as described previously (algorithm 51).

4 Notes

Single-cell multi-omics sequencing technologies have revolutionized the study of tumor heterogeneity by providing high-resolution insights into the identities and functional states of tumor cells. These techniques, such as scRNA-seq and scDNA-seq, enable precise differentiation between malignant and nonmalignant cells, revealing critical molecular markers and genomic alterations. By integrating multiple layers of molecular data—such as transcriptome, DNA methylation, mutational profiles, and chromatin structure—these approaches offer a more nuanced view of tumor biology than single-omic methods. Web-based multi-omics tools enable investigation of the genetic and non-genetic drivers of cancer progression, capturing intricate regulatory interactions between various cellular states and molecular processes.

Despite these advancements, significant challenges remain in translating multi-omics insights into clinical applications. Data integration is a major hurdle due to the heterogeneity of datasets, varying measurement scales, and the complexity of aligning multimodal information. Additionally, the high cost, technical noise, and low-throughput nature of current methods limit their application to large-scale studies, particularly in complex tissues with rare cell populations. As sequencing technologies advance and costs decrease, more scalable computational solutions, including artificial intelligence (AI)-enabled machine learning and cloud infrastructure, will be required to process and analyze the expanding datasets. Overcoming these challenges will be crucial for fully harnessing the potential of single-cell multi-omics technologies, ultimately driving progress in precision medicine and improving disease diagnosis, treatment, and outcomes.

Conflict of Interest

The authors have no conflicts of interest to declare that are relevant to the content of this chapter.

References

1. Chakraborty S et al (2024) Multi-OMICS approaches in cancer biology: new era in cancer therapy. Biochim Biophys Acta Mol basis Dis 1870(5):167120
2. Sun Y et al (2023) Single-cell multi-omics sequencing and its application in tumor heterogeneity. Brief Funct Genomics 22(4):313–328
3. Akhoundova D, Rubin MA (2022) Clinical application of advanced multi-omics tumor profiling: shaping precision oncology of the future. Cancer Cell 40(9):920–938
4. Nam AS, Chaligne R, Landau DA (2021) Integrating genetic and non-genetic determinants of cancer evolution by single-cell multi-omics. Nat Rev Genet 22(1):3–18
5. Zhou G et al (2022) OmicsNet 2.0: a web-based platform for multi-omics

integration and network visual analytics. Nucleic Acids Res 50(W1):W527–W533
6. Ewald JD et al (2024) Web-based multi-omics integration using the analyst software suite. Nat Protoc 19(5):1467–1497
7. Wang J et al (2018) Medulloblastoma: from molecular subgroups to molecular targeted therapies. Annu Rev Neurosci 41:207–232
8. Cavalli FMG et al (2017) Intertumoral heterogeneity within Medulloblastoma subgroups. Cancer Cell 31(6):737–754.e6
9. Luo Z et al (2022) Human fetal cerebellar cell atlas informs medulloblastoma origin and oncogenesis. Nature 612(7941):787–794
10. Luo Z et al (2025) Multimodal validation of the existence of transitional cerebellar progenitors in the human fetal cerebellum. Preprint at *bioRxiv*. https://doi.org/10.1101/2025.06.01.657310
11. McGinnis CS, Murrow LM, Gartner ZJ (2019) DoubletFinder: doublet detection in single-cell RNA sequencing data using artificial nearest neighbors. Cell Syst 8(4):329–337 e4
12. Liu M et al (2021) HNRNPH1 is a novel regulator of cellular proliferation and disease progression in chronic myeloid leukemia. Front Oncol 11:682859
13. Oliemuller E et al (2020) SOX11 promotes epithelial/mesenchymal hybrid state and alters tropism of invasive breast cancer cells. elife 9:9
14. Patmore DM et al (2020) DDX3X suppresses the susceptibility of hindbrain lineages to Medulloblastoma. Dev Cell 54(4):455–470 e5
15. Muskovic W, Powell JE (2021) DropletQC: improved identification of empty droplets and damaged cells in single-cell RNA-seq data. Genome Biol 22(1):329
16. Qiu X et al (2017) Reversed graph embedding resolves complex single-cell trajectories. Nat Methods 14(10):979–982
17. Street K et al (2018) Slingshot: cell lineage and pseudotime inference for single-cell transcriptomics. BMC Genomics 19(1):477
18. Chen H et al (2019) Single-cell trajectories reconstruction, exploration and mapping of omics data with STREAM. Nat Commun 10(1):1903
19. Zhang F, Li X, Tian W (2020) Unsupervised inference of developmental directions for single cells using VECTOR. Cell Rep 32(8):108069
20. Setty M et al (2016) Wishbone identifies bifurcating developmental trajectories from single-cell data. Nat Biotechnol 34(6):637–645
21. Trapnell C et al (2014) The dynamics and regulators of cell fate decisions are revealed by pseudotemporal ordering of single cells. Nat Biotechnol 32(4):381–386
22. Steen CB et al (2020) Profiling cell type abundance and expression in bulk tissues with CIBERSORTx. Methods Mol Biol 2117:135–157
23. Hovestadt V et al (2019) Resolving medulloblastoma cellular architecture by single-cell genomics. Nature 572(7767):74–79
24. Northcott PA et al (2017) The whole-genome landscape of medulloblastoma subtypes. Nature 547(7663):311–317
25. Neftel C et al (2019) An integrative model of cellular states, plasticity, and genetics for glioblastoma. Cell 178(4):835–849 e21
26. Jessa S et al (2019) Stalled developmental programs at the root of pediatric brain tumors. Nat Genet 51(12):1702–1713
27. Tirosh I et al (2016) Single-cell RNA-seq supports a developmental hierarchy in human oligodendroglioma. Nature 539(7628):309–313
28. Granja JM et al (2021) ArchR is a scalable software package for integrative single-cell chromatin accessibility analysis. Nat Genet 53(3):403–411
29. Corces MR et al (2018) The chromatin accessibility landscape of primary human cancers. Science 362(6413)
30. Aibar S et al (2017) SCENIC: single-cell regulatory network inference and clustering. Nat Methods 14(11):1083–1086
31. Wang X et al (2021) Genome-wide detection of enhancer-hijacking events from chromatin interaction data in rearranged genomes. Nat Methods 18(6):661–668
32. Riecken LB et al (2016) CPI-17 drives oncogenic Ras signaling in human melanomas via Ezrin-radixin-Moesin family proteins. Oncotarget 7(48):78242–78254

Chapter 2

Derivation of Primary Cell Lines from Patient Tumor Samples for In Vitro and In Vivo Studies of Malignant Gliomas

Rowland H. Han, Diane D. Mao, Timothy R. Woodiwiss, Zelun Wang, and Albert H. Kim

Abstract

Primary glioblastoma (GBM) cell lines can be derived from surgically resected patient tumor specimens. The main advantage of these lines is that they preserve the genetic, epigenetic, and metabolic features of the original tumor. These cell lines can be used for a variety of in vitro assays and orthotopically implanted into immunodeficient mice for in vivo experiments. We describe a method for isolation, maintenance, propagation, and in vivo injection of patient-derived GBM sphere cell lines.

Key words Glioblastoma, Glioblastoma stem cells, Patient-derived culture, Sphere culture

1 Introduction

Primary glioblastoma (GBM) cell lines can be isolated from surgically resected human tumor specimens and propagated in sphere culture or as adherent cell lines [1–5]. These cell lines retain key features of the parental tumor, represent a valuable tool to study the biology underlying tumor behavior, and are able to form tumors in vivo when injected orthotopically into immunodeficient mice [6, 7]. Primary GBM lines grown in culture are one of the few reliable methods to study living GBM cells from human tumor specimens, reflecting the intra- and inter-tumoral genetic heterogeneity that exists in GBM.

Our laboratory routinely generates sphere culture cell lines from freshly resected tumors using methods modified from other groups [8, 9] and our own experience, using mechanical and enzymatic dissociation of tumor specimens to prepare a homogeneous suspension [10–12]. The spheroid growths are then propagated in suspension with serum-free medium, supplemented with epidermal

Michael Goldstein (ed.), *Translational Research Methods for Pediatric Brain Tumors*, Neuromethods, vol. 226,
https://doi.org/10.1007/978-1-0716-4754-7_2,

growth factor (EGF) and fibroblast growth factor (FGF). We describe these and additional protocols for freezing and thawing the cells. Finally, we describe our procedure for orthotopic implantation of the cell lines in immunodeficient mice for in vivo experimentation.

2 Materials

2.1 Equipment

1. Sterile ultralow attachment tissue culture dishes (60 mm or 100 mm, Corning).
2. Sterile conical tubes (15 and 50 ml).
3. Benchtop centrifuge.
4. Water bath.
5. Humidified tissue culture incubator (37 °C, 5% CO_2).
6. 70-μm cell strainer.
7. Cryogenic vials (for freezing cells).
8. Cell freezing containers.
9. Cell counter.
10. Stereotactic apparatus and micro-injector.
11. Hamilton syringe.
12. Standard sterile instruments including scalpel and forceps.
13. Isoflurane and prep solutions for animal surgery.

2.2 Solutions and Buffers

1. Ringer's solution.
2. Tissue culture-grade water.
3. Phosphate-buffered saline (PBS).
4. Hanks' Balanced Salt Solution (HBSS).
5. Accutase cell dissociation solution (Gibco #A11105-01). Alternative enzymatic solutions are Accumax or TrypLE (*see* **Note 1**).
6. Dulbecco's Modification of Eagle's Medium / Ham's F-12 (DMEM/F-12) with GlutaMAX (1X) (Gibco #10565-018).
7. Freezing medium: 85% DMEM/F-12, 5% bovine serum albumin solution, 10% DMSO.

2.3 Reagents and Supplements

1. Epidermal growth factor (EGF) (human) (PeproTech #AF-100-15).
2. Fibroblast growth factor-Basic (FGF) (human) (PeproTech #100-18B).
3. B-27 supplement without Vitamin A (50X), serum free (Miltenyi Biotec #130-097-263).

4. Penicillin-Streptomycin (10,000 U/mL) (ThermoFisher #15140122).
5. Heparin sodium salt from porcine intestinal mucosa (Sigma-Aldrich #H3149-50KU).
6. Dimethyl sulfoxide (DMSO), cell-culture grade.
7. Bovine serum albumin, cell-culture grade.
8. Ammonium-Chloride-Potassium (ACK) lysing buffer (Gibco #A10492–01).

3 Methods

3.1 Preparation of GBM Sphere Culture Media

1. Open a fresh 500 ml bottle of DMEM/F12 with GlutaMAX. Add 1.25 ml of Penicillin-Streptomycin and 10 ml of B-27 supplement. This incomplete medium can be stored at 4 °C for up to 1 month.
2. For complete medium, take a 50 ml aliquot of incomplete medium. Add 50 ml heparin (final concentration 2.5 μg/ml), 10 μl EGF (final concentration 20 ng/ml), and 10 μl FGF (final concentration 20 ng/ml) (*see* **Note 2**). Growth factors degrade quickly in medium (2–3 days), so heparin, EGF, and FGF should be added fresh as needed.

3.2 Specimen Dissociation and Establishment of Culture

1. After surgical resection, keep the tumor specimen in cold Ringer's solution and process as soon as possible. We typically process within 30 min of surgical resection.
2. Transfer the specimen to any tissue culture-grade sterile petri dish and rinse once with PBS. Remove PBS.
3. Mechanical dissociation: In the dish, mince the specimen well using a clean scalpel until it can pass through a 10 ml serological pipette (*see* **Note 3**).
4. Enzymatic dissociation: Add 2–3 ml of Accutase solution to the dish and pipette up and down several times using a 10 ml pipette. Transfer the minced specimen with Accutase to a sterile conical tube and allow chemical dissociation for 20–60 min at 37 °C in a water bath. Occasionally tap or invert gently to resuspend/mix.
5. After incubation, stop dissociation by adding 5 ml of prewarmed complete sphere culture medium to the tube, mix gently with pipette, and centrifuge at 300× *g* for 3 min. Discard the supernatant.
6. Resuspend the pellet in 2–3 ml of fresh prewarmed complete sphere culture medium, and make a homogeneous suspension by pipetting up and down using a 5 ml serological pipette or a 1 ml pipette tip with a P1000 pipettor.

7. Filter the suspension through a 70-μm cell strainer into a 50 ml conical tube.
8. Centrifuge the suspension at 300× *g* for 3 min. Discard the supernatant.
9. If the cell pellet appears red (contains red blood cells), add 1–2 ml ACK lysing buffer. Lyse the pellet at room temperature for 2 min, then add 10–20 ml HBSS to neutralize the red blood cell lysing buffer. Centrifuge the cells at 300× *g* for 3 min and discard the supernatant. This step can be repeated if the pellet still appears red. Once red blood cells have been removed, add 5 ml of complete sphere culture medium (*see* **Note 4**).
10. Wash the cells an additional two times by centrifugation and resuspension in complete sphere culture medium.
11. Resuspend the final cell pellet in fresh prewarmed complete sphere culture medium and plate cells in an ultralow attachment dish (*see* **Note 5**). Store dishes in an incubator at 37 °C and 5% CO_2.

3.3 Maintenance and Passage of Cultures

1. Check the culture dishes daily for growth of spheres (*see* **Note 6**). When the centers of the spheres become dark, typically around 150–250 μm in diameter (Fig. 1), the cells should be passaged as follows:
2. Collect the cell suspension into a 15 ml conical tube, centrifuge at 200× g for 5 min, and keep the media in the supernatant if not yellow. This conditioned medium can be mixed with new sphere medium (plus fresh growth factors and heparin for the entire volume) when resuspending cells for passage. This can also save on overall media costs.

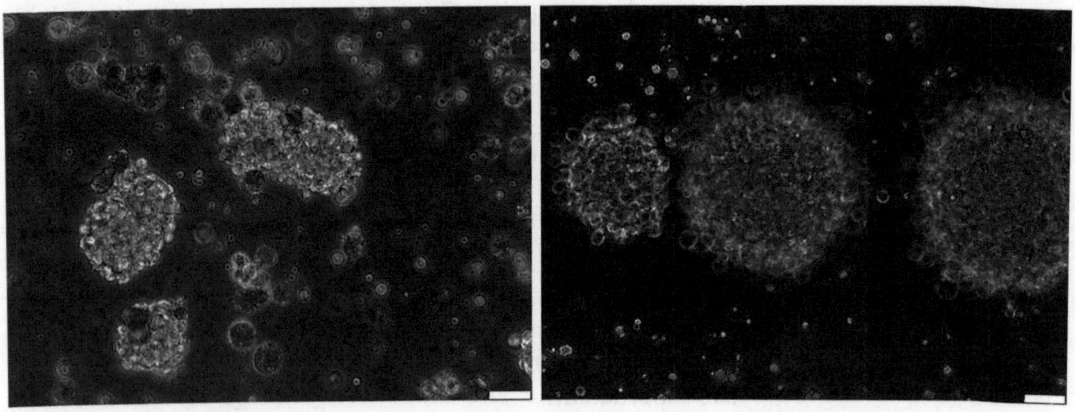

Fig. 1 Photomicrographs of the B172 cell line in sphere culture when actively growing (left, note brightness of the cells) and when spheres become dark and, thus ready to passage (right). Scale bars represent 50 μm

3. Add 1 ml of Accutase to the pellet and pipette to resuspend. Incubate with Accutase at 37 °C for 5–7 min, or until spheres break apart (*see* **Note 1**).
4. Add 3–5 ml of complete sphere culture medium to stop the digestion. Centrifuge at 200× *g* for 5 min, and discard the supernatant.
5. Resuspend cells in fresh complete sphere culture medium in an ultralow attachment dish at 1:3 to 1:5 dilution and observe under a microscope. The suspension should now contain single cells.
6. If needed, the cells can be counted using a hemocytometer or automatic cell counter and used for assays or immediate in vivo implantation instead of replating in fresh medium.
7. After expansion to passage 5, we freeze cells for liquid nitrogen storage for subsequent routine use in experiments (*see* **Note 7**).

3.4 Freezing Cells

1. Follow **steps 2–4** in Subheading 3.3.
2. Resuspend the cell pellet in freezing medium at the desired dilution. We generally freeze at 0.5–1 × 10^6 cells/ml.
3. Prepare 1 ml aliquots of the suspension in sterile cryogenic vials.
4. Store the vials in a cell freezing container at −80 °C and transfer to liquid nitrogen after 2–3 days.

3.5 Thawing Cells

1. Take the previously stored cell vials from liquid nitrogen and quickly thaw in a 37 °C water bath with continuous gentle agitation.
2. When partially thawed, immediately transfer the cells from the vial to a 15 ml conical tube containing 9 ml of prewarmed sphere culture medium.
3. Centrifuge at 200× *g* for 5 min.
4. Discard the supernatant and resuspend the cell pellet in fresh prewarmed complete sphere culture medium.
5. Transfer the cells to an ultralow attachment dish.
6. Passage cells 1–2 times after thawing to ensure they are growing well before using them for in vitro or in vivo experiments.

3.6 Implantation of Primary Cell Lines in Mice for In Vivo Experiments

1. Patient-derived primary GBM cell lines can be implanted in 6- to 9-week-old athymic nude (Crl:NU(NCr)-*Foxn1*^{nu}, Charles River) mice using standard stereotactic injection procedures as detailed below (Fig. 2). NOD SCID or NOD SCID gamma mice may also be used if desired.
2. Follow **steps 2–4** in Subheading 3.3 to harvest healthy, growing single cells.

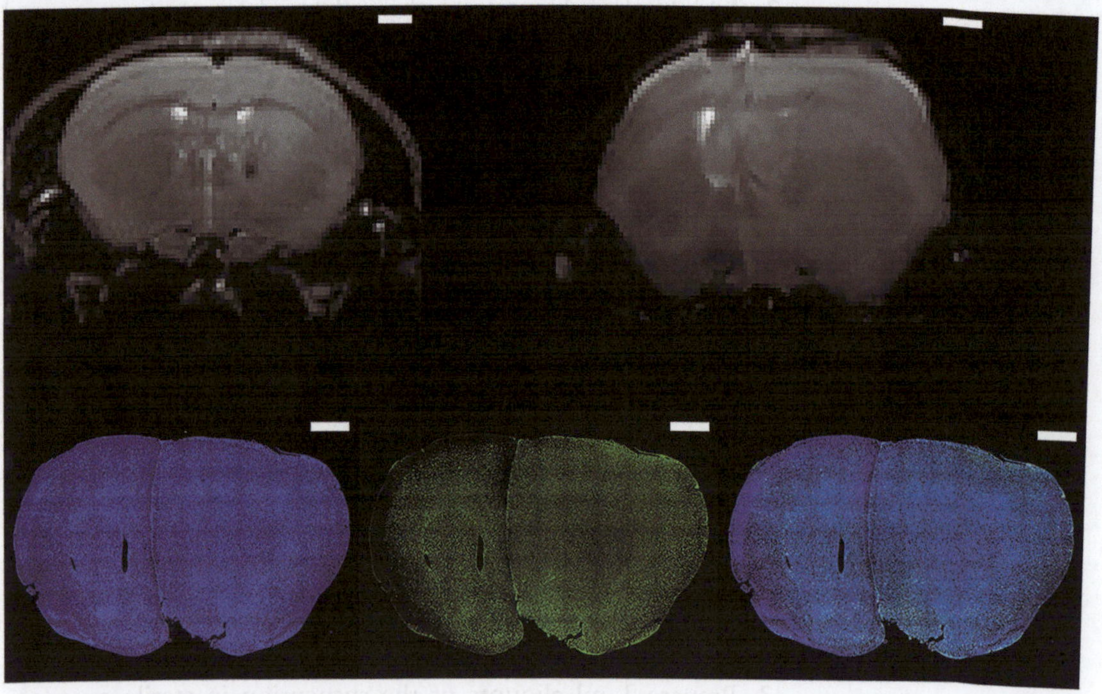

Fig. 2 Patient-derived GBM cell lines form tumors when implanted into immunodeficient mice. Top: The B172 cell line was injected into an athymic nude mouse which underwent T2-weighted MRI at 8 weeks (left) and 16 weeks (right) post-implantation. The T2-hyperintense region primarily on the right side represents the tumor. Bottom: The brain from the same mouse was harvested at 16 weeks postimplantation and immunofluorescence performed on formalin-fixed paraffin-embedded coronal sections with anti-human nucleoli antibody [NM95, Abcam, ab190710] and Hoechst counterstain (blue). Immunofluorescence shows deeply infiltrative human tumor cells (green) throughout both hemispheres. Scale bars represent 1 mm

3. Resuspend cells in 1 ml of ice-cold PBS and count live cells using a hemocytometer or automatic cell counter. It is best to prepare 2–3 times the number of cells required to inject the desired number of mice, to ensure all mice can be injected.
4. Centrifuge cells in PBS at 200× *g* for 5 min, then discard the supernatant.
5. Resuspend cells in fresh ice-cold PBS at the desired concentration for injections based on the results of cell counting. We typically use 2.5×10^5 cells in 2ul of PBS per mouse for patient-derived lines. Keep this final suspension on ice until ready to inject.
6. Place an adequately anesthetized mouse on the stereotactic apparatus. Using standard sterile techniques, incise the skin, and navigate the syringe to the desired injection coordinates. We typically use ML 2.5 mm (right), AP 1 mm (anterior), DV 2.5 mm (deep) from the bregma.
7. Drill a burr hole in the skull at the desired location.

8. Load the microsyringe with 2 μl of cell suspension.
9. Insert the microsyringe to a depth of 3 mm and let sit for 30 seconds, then retract it by 0.5 mm, creating a pocket with the syringe at an injection depth of 2.5 mm. Wait 1–2 min.
10. Slowly inject the full 2 μl of the cell suspension over 5 min. We use an automatic micro-injector to help with this step. Wait 1 min after the injection is complete to avoid reflux up the needle tract.
11. Retract the microsyringe by 0.5 mm, then wait 1 minute. Repeat until the microsyringe is fully removed from the brain.
12. Gently wipe the injection site with a cotton-tipped applicator dabbed in a very small amount of 70% ethanol to remove any cells at the surface.
13. Using the reverse end of the applicator, apply a small amount of bone wax or glue into the burr hole to prevent reflux of cells.
14. Close the incision with staples or suture. Remove on postimplantation day 7–10.

4 Notes

1. Some cell lines may require stronger enzymes than Accutase for cell dissociation. In these cases, Accumax (ThermoFisher #00-4666-56) or TrypLE (ThermoFisher #12605010) may be used. However, these stronger enzymes may result in more cell death. When using TrypLE, do not digest for more than 2 min.
2. Store stock solutions of EGF and FGF in 50 μl aliquots at a concentration of 100 μg/ml at −80 °C. Aliquots can be used up to 7 days after thawing and added to incomplete sphere culture medium just before use.
3. Gentle but thorough mechanical dissociation of the tumor specimen is important to obtain a homogeneous cell suspension. This can be achieved by gently mincing the specimen with a scalpel for 3–5 min until it has a jelly-like consistency.
4. Lympholyte-M (Cedarlane) is another method for the removal of red blood cells, dead cells, and debris.
5. We have found that primary cell lines grow best when using 60 mm ultralow attachment dishes. The cells can later be moved to 100 mm dishes if they are growing rapidly.
6. It is important to check the cells often to make sure they are not depleting the nutrients in the media too quickly. Some lines grow slowly and may not need to be passaged before the media begins to change from pink to yellow. In this case, fresh EGF,

FGF, and heparin should be added every 2–3 days to the suspension without removing media. It may take several days (range 7–30 days) for spheres to grow and darken for the first passage.

7. In most cases, it is not advisable to use cells after passage 30 to avoid genetic alterations that may be associated with repeated passages. We also find that biological phenotypes can change after passage numbers in the low 20 s.

References

1. Caren H et al (2015) Glioblastoma stem cells respond to differentiation cues but fail to undergo commitment and terminal cell-cycle arrest. Stem Cell Rep 5(5):829–842
2. Pollard SM et al (2009) Glioma stem cell lines expanded in adherent culture have tumor-specific phenotypes and are suitable for chemical and genetic screens. Cell Stem Cell 4(6): 568–580
3. Singh SK et al (2004) Identification of human brain tumour initiating cells. Nature 432(7015):396–401
4. Chen J et al (2012) A restricted cell population propagates glioblastoma growth after chemotherapy. Nature 488(7412):522–526
5. Bao S et al (2006) Glioma stem cells promote radioresistance by preferential activation of the DNA damage response. Nature 444(7120): 756–760
6. Fael Al-Mayhani TM et al (2009) An efficient method for derivation and propagation of glioblastoma cell lines that conserves the molecular profile of their original tumours. J Neurosci Methods 176(2):192–199
7. Lee J et al (2006) Tumor stem cells derived from glioblastomas cultured in bFGF and EGF more closely mirror the phenotype and genotype of primary tumors than do serum-cultured cell lines. Cancer Cell 9(5):391–403
8. Mao P et al (2013) Mesenchymal glioma stem cells are maintained by activated glycolytic metabolism involving aldehyde dehydrogenase 1A3. Proc Natl Acad Sci USA 110(21): 8644–8649
9. Bhat KPL et al (2013) Mesenchymal differentiation mediated by NF-kappaB promotes radiation resistance in glioblastoma. Cancer Cell 24(3):331–346
10. Mao DD et al (2015) A CDC20-APC/SOX2 signaling Axis regulates human glioblastoma stem-like cells. Cell Rep 11(11):1809–1821
11. Mahlokozera T et al (2021) Competitive binding of E3 ligases TRIM26 and WWP2 controls SOX2 in glioblastoma. Nat Commun 12(1): 6321
12. Gujar AD et al (2018) Establishing primary human glioblastoma adherent cultures from operative specimens. Methods Mol Biol 1741: 53–62

Chapter 3

In Vivo Murine Model for Oncolytic Viral Therapy for Malignant Glioma

Ashwani Kesarwani, Yuping Derek Li, Tong Hu, and Milan G. Chheda

Abstract

Glioblastoma (GBM) is the most common and aggressive primary brain tumor. Even with recent developments in immunotherapy, the prognosis for GBM remains poor, and tumor recurrence is inevitable. Consequently, preclinical models for testing new therapies for GBM are critical for developing better treatments. Specifically, for investigating oncolytic viruses, a syngeneic tumor model is necessary so that an immunocompetent host can be used. Here, we describe our protocol for an in vivo murine model to test the efficacy of oncolytic *Zika virus* to treat high grade gliomas. These methods can be expanded to incorporate other tumor cell lines as well as other oncolytic viruses.

Key words Oncolytic viruses, Zika virus, Glioblastoma, GL261, CT2A, SB28

1 Introduction

1.1 Brain Tumors and Glioblastoma

Glioblastoma (GBM) is an aggressive and rapidly growing brain tumor affecting children, adolescents, and adults that invades neighboring brain tissue but rarely metastasizes. According to the World Health Organization (WHO) classification [1, 2]. GBM is a grade 4, IDH wild-type astrocytoma, a tumor that arises from (astro-) glial cells and their progenitors [3]; however, the specific cell of origin of GBM is unclear. Clinically, the prognosis is poor, largely due to inevitable tumor recurrence. Survival without therapy is 4–7 months. Following surgical resection with adjuvant radiation, temozolomide, and tumor-treating fields, the median survival for GBM patients is less than 2 years. Five-year survival is only three percent, and 10-year survival is one percent [4, 5].

GBM is associated with several hereditary illnesses, including neurofibromatosis [6], tuberous sclerosis, and Li-Fraumeni syndrome. There is a higher incidence of GBM in males [7–9]. Persistent headaches, nausea, vomiting, impaired vision, changes in cognitive ability, memory loss, personality changes, seizures,

Michael Goldstein (ed.), *Translational Research Methods for Pediatric Brain Tumors*, Neuromethods, vol. 226,
https://doi.org/10.1007/978-1-0716-4754-7_3,

muscular weakness, and/or trouble speaking are all symptoms of GBM. Magnetic resonance imaging (MRI) and, to a lesser extent, computed tomography (CT) scan play an important diagnostic role. Pathologic assessment of the tumor specimen is required for the diagnosis. There are several barriers for effective treatment for GBM, including the following: (1) the tumor microenvironment (TME) has a paucity of T cells [10, 11] and, consequently, immunotherapies have failed in multiple phase III clinical trials [12, 13], and (2) a treatment resistant population of cancer stem cells contributes to inevitable tumor recurrence [14–16]. Here, we review several common experimental models of glioma and the rationale and use of oncolytic *Zika virus* to understand brain tumor biology.

1.2 Glioma Mouse Models

Glioma induction can be achieved in at least six ways: (1) human glioma cell xenografts implanted into immunocompromised mice directly from the operating room or only briefly passaged in culture (patient-derived xenograft, PDX) [17–21], (2) human GBM cell line xenografts transplanted into immunocompromised mice [19, 22], (3) human DR4+ hematopoietic stem cells transplantation to generate a humanized DRAG (NOD.Rag1KO.IL2RγcKO) mouse model and use of GBM patient-derived tumorsphere cells to form xenografted intracranial tumors [23], (4) induction of autochthonous tumors in genetically modified mice (genetically engineered model), (5) CRISPR-based in utero electroporation (IUE) to characterize a mouse model of GBM in immunocompetent mice [24], or (6) syngeneic tumor cell implantation in immunocompetent host mice (syngeneic model). Several syngeneic mouse models exist that may lead to a better understanding of the immunological interactions within the tumor microenvironment. The most significant benefits of syngeneic models are the preserved immunocompetence of the host, recapitulation of the biology and histological aspects of high-grade gliomas, the degree of control of location, the reproducible timing of tumor development, and the relatively low cost (as opposed to breeding genetically modified mice) [25]. However, syngeneic models also have downsides, such as the absence of a native tumor microenvironment, limited host strains, relatively few transplantable cell lines, and the lack of the complex heterogeneity that exists in patients [26]. Murine models created via PDX or genetically modified procedures might be useful in these cases. Because they preserve both the histological and genetic properties of the originating tumor, PDX models are often employed in translational research investigations. Genetically modified models, on the other hand, are frequently employed for evaluating treatment techniques because they allow researchers to identify genetic changes implicated in tumor pathogenesis [20, 25]. The following are the most common syngeneic mouse models:

1.2.1 GL261

Seligman and colleagues created the GL261 murine model in 1970 by implanting a methylcholanthrene (MCA) pellet intracranially. Following that, Ausman et al. further induced tumor via intracranial injection of MCA followed by serial intracranial and subcutaneous transplantations of tumor fragments (1×10^5 tumor cells/10 ul) directly into the brains of C57BL/6 mice, achieving 99.1% tumor formation, median survival of 22.1 to 25.9 days, and histologic stability for more than 100 generations [27]. The GL261 model has been one of the most widely used murine glioma models since its inception. In recent years, with the advent of immunotherapies and the focus on the immunological microenvironment, GL261 has come under more scrutiny. Its relatively high mutational burden differs from the low mutational burden more typical of GBM. In addition, it does not phenocopy the immunosuppressive microenvironment of human brain tumors [28, 29]. A high mutational burden increases the chances that a tumor may respond to immune checkpoint blockade treatment, such as anti-PD-1 and CTLA4 therapy [30–33]. Different than the human situation, GL261 is highly responsive to immune checkpoint blockade [28, 34, 35]. In summary, GL261 cells serve as a model system for high-grade astrocytoma, including glioblastomas (GBMs). These cells offer valuable insights into glioma stem cells, which play a role in the chemo- and radioresistance observed in human GBM. Researchers can utilize this model to assess novel therapeutic strategies.

1.2.2 SB28

Because human GBM has limited immunogenicity and an immunosuppressive tumor microenvironment, adopting a comparably less immunogenic mouse model is helpful in determining the effectiveness of immunotherapeutic drugs [36]. The SB28 cell line was created by transfecting the Sleeping Beauty (SB) transposon-flanked proto-oncogene into a newborn C57BL/6 mouse [37, 38]. The model is driven by transfection of transgenes like P53, PDGFB, and NRAS, which are introduced into the cells using the **Sleeping Beauty transposon system**. It has a molecular signaling similarity to the driver mutations often observed in the proneural group of GBM [37–39]. SB28 has recently been investigated to determine its similarities and distinctions compared to more well-established models such as GL261; however, there is still not enough information addressing its immunological profile [28]. SB28 gliomas exhibit hypervascularization and high cellularity inside the tumor boundaries, as well as invasion of normal surrounding brain parenchyma. Tumor induction is reliable and successful for SB28 gliomas, with a 100% tumor induction rate and a median survival of 29 days in mice bearing SB28. This month's timeline provides enough time for research of therapeutic perturbations and assessment [28]. SB28 gliomas have a lower mutational burden than GL261, allowing for a more faithful approximation of human GBM as well as a more accurate response to prospective immunotherapeutic drugs in preclinical studies [28]. In summary, SB28 is a weakly immunogenic mouse glioma model that strives to accurately replicate the tumor immunogenicity and microenvironment observed in human GBM. This distinctive characteristic enables researchers to investigate GBM progression and therapeutic responses with greater precision.

Unlike GL261, which is highly immunogenic, SB28 has a more attenuated immune profile, making it better for modeling the immunosuppressive human GBM microenvironment [28]. In addition, MHC I expression is intimately related to the tumor cell's innate immunogenicity. SB28, unlike GL261, is constitutively MHC I negative. Interferon-γ stimulation results in upregulation of MHC-I, but does not induce MHC II expression in SB28 [28, 40]. Furthermore, treatment of SB28 tumors with dual anti-PD1 and anti-CTLA4 immune checkpoint inhibitors does not improve survival compared to 50% tumor clearance of GL261 tumors [28, 41–43].

CT2A: CT2A was created by implanting MCA pellets intracranially in immunocompetent C57BL/6 mice. The CT2A cell line faithfully produces brain tumors when implanted into mice, and mice die between 3 to 8 weeks of tumor growth [44]. CT2A gliomas exhibit histological characteristics such as high cellular density, pseudo-pallisading necrosis, and irregularly shaped cells with high and aberrant mitotic activity [44, 45]. Like many human GBMs, CT2A cells express p53, GFAP, and Sox9 [46, 47]. Also, similar to the human context, CT2A tumors are highly proliferative and infiltrative [48, 49]. CT2A cells share characteristics with CD133+ glioma stem cells, which are capable of recurrent tumor formation [44, 49]. In humans, studies have found that GBMs with a higher expression of CD133 are associated with worse progression-free and overall survival [50]. When grown in serum-free conditions, CT2A cells form neurospheres, which when implanted into mice provide a more aggressive tumor, both histologically and immunologically than CT2A cells, which were grown in standard media with serum [51]. CT2A has a genetic mutational burden that lies between SB28 and GL261, which is relatively higher than that observed in human GBM [28, 52]. While CT2A gliomas may not exactly mimic the human GBM phenotype, they are a useful preclinical mouse models due to their comparable histological, immunohistochemical, proliferative, and metabolic characteristics [44, 47]. CT2A gliomas have higher levels of immune checkpoint molecule (PD-L1) expression [44, 53, 54] and are also more resistant to checkpoint blockade treatment than GL261 gliomas [52, 53]. Anti-PD-1 treatment resulted in a small number of long-term CT2A glioma survivors, which was increased by the inclusion of radiation. In contrast to the 50%

long-term survival rates found in GL261-bearing mice [34], CT2A-bearing mice treated with anti-PD-L1 exhibited marginally better median survival (4.5 days) [52]. Oncolytic viral treatment has also been tested in the CT2A model, with promising results. Barnard et al. found that oncolytic herpes simplex virus 1 vectors expressing FMS-like tyrosine kinase 3 ligand (Flt3L) increased survival in CT2A mice [55]. Within the glioma, treated mice had more immature dendritic cells, cytotoxic T lymphocytes, and cytokine expression. To summarize, CT2A is a weakly immunogenic mouse model that is very helpful in understanding the role of glioma stem cells in glioma growth and the tumor microenvironment. We recommend the use of multiple models when testing oncolytic viruses.

1.3 Oncolytic Viruses

Viruses are small particulate entities, which possess the capability to replicate within host cells, eliciting an inflammatory response in the host. They are primarily categorized into DNA and RNA viruses [56]. DNA viruses, like adenoviruses, are characterized by their extensive genomes, which can be manipulated to incorporate transgenes that augment therapeutic efficacy through immune modulation or alternative pathways without compromising viral propagation [57]. Conversely, RNA viruses, like Zika virus, exhibit a more compact genome compared to their DNA counterparts and are adept at traversing the blood-brain barrier (BBB) to target neoplastic cells within the central nervous system (CNS) [58]. The constrained genomic capacity of RNA viruses restricts their potential for encoding expansive transgenes. Furthermore, they exhibit heightened genetic instability and elevated mutation frequencies relative to DNA viruses [59].

Over the past decade, both natural and engineered oncolytic viruses (OVs) derived from DNA and RNA viruses have been experimentally tested [60, 61], with some advancing to clinical trials. Notable OVs currently in clinical trials include herpes simplex virus (rQNestin34.5v.2) (Clinical Trial: NCT03152318), also for glioma patients [62]; measles virus (Clinical Trial: NCT02068794), for drug-resistant ovarian cancer [63]; adenovirus (E1B-attenuated oAd, ONYX-015) (Clinical Trial: NCT03178032), for glioma patients [64]; and vaccinia virus (Clinical Trial: NCT01469611), for colorectal cancer [65]; and DNX-2401 (Delta-24-RGD) adenovirus (Clinical Trial: NCT00805376), for recurrent malignant gliomas [66, 67]. Additionally, T-VEC (HSV1) is FDA approved for treating unresectable stage IIIB-IV melanoma [68]. We have pursued *Zika virus* as a possible glioma therapy [69, 70].

Oncolytic viruses employ several mechanisms to eliminate tumor cells:

Direct lysis: The virus infects and replicates within the cancer cell, leading to cell rupture and the release of additional infectious viral particles and tumor antigens [71].

Vascular disruption: The virus can infect and destroy blood vessels supplying the tumor, thereby limiting oxygen and nutrient delivery [72]. Additionally, the virus may induce immunogenic cell death, a process that enhances antitumor immunity [73, 74].

Gene therapy: These viruses have the capacity to transfer therapeutic genes to tumor cells, including genes that enhance cellular susceptibility to chemotherapy or radiation, or genes that promote cell death or differentiation [75]. These intricate mechanisms contribute to the efficacy of oncolytic viruses in combating cancer [76, 77].

2 Materials

2.1 Mice

1. C57BL/6 J mice.
2. It is important to implant tumors in 10–20% extra mice needed for downstream studies, to account for experimental variability in tumor engraftment rate.

2.2 Cell Preparation

1. 10 cm tissue culture dishes.
2. CT2A media (DMEM media +10% FBS + 1% penicillin-streptomycin).
3. GL261 media (RPMI +10% FBS + 1% penicillin-streptomycin 1% L-glutamine).
4. SB28 media (RPMI +10% FBS + 1% penicillin-streptomycin +1% L-glutamine +1% sodium pyruvate +1% HEPES +1% non-essential amino acids).
5. Phosphate-buffered saline without Ca^{+2} or Mg^{+2}.
6. 0.05% trypsin.
7. Trypan blue.
8. Hemacytometer.

2.3 Equipment

1. Stereotactic platform for mouse (Stoelting Inc.) (Fig. 2).
2. Micro Drill with 0.45 mm drill bit (Stoelting Inc.).
3. Hamilton syringe, 26-gauge.
4. Sterile #10 blade scalpel.
5. Ketamine–xylazine anesthetic solution (10 mg/mL and 1 mg/mL, respectively).
6. Buprenorphine–ER analgesic solution (1 mg/mL).
7. 29G insulin syringes.

8. Small animal hair clipper.
9. 10% Betadine solution (antiseptic solution).
10. Sterile lubricant eye ointment (Stye).
11. Alcohol swabs.
12. Fine black marker to mark site of burr hole for drilling.
13. Sterile cotton swabs.
14. 4–0 monofilament nylon suture (McKesson).
15. Vetbond tissue adhesive (3 M).
16. Heating pad.
17. D-luciferin (GoldBio) (15 mg/mL).

3 Methods

3.1 Orthotopic Tumor Implantation

Preimplantation tumor cell preparation

1. Perform standard cell culture using CT2A, GL261, or SB28 cells in their respective media. Cells are grown in 10 cm tissue culture dishes incubating at 5% CO_2 and 37 °C (Fig. 1, **step 1**).

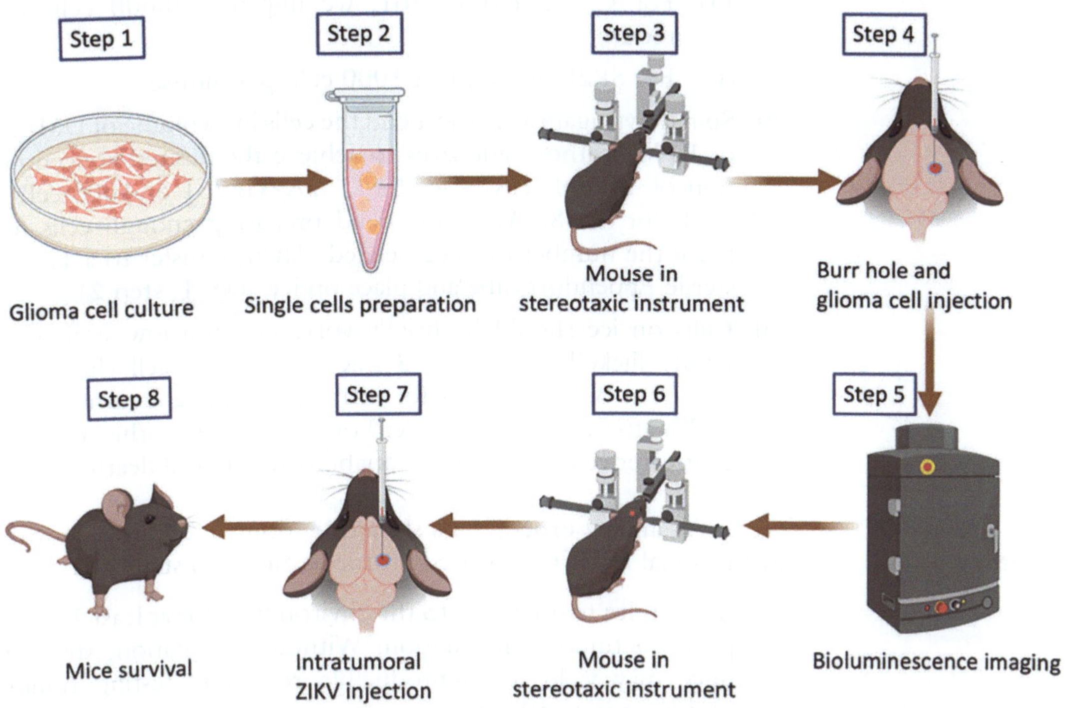

Fig. 1 Stepwise procedure for the preparation and implantation of glioma cells in the mouse brain

2. The CT2A, GL261, and SB28 cell lines used in our lab express the firefly luciferase gene. If the cells being used have not been labelled for bioluminescence, transduce each cell line with an expression vector to stably express the firefly luciferase gene.
3. Prior to implantation, cells should be passaged two to three times after being thawed; this allows for optimal tumor engraftment and growth in vivo.
4. On day of implantation, cells should be approximately 70–80% confluent, so that they are in log-phase growth at time of implantation. Remove media and begin by washing the cells with 1X phosphate buffered saline (PBS) followed by 0.05% trypsin (pre-warmed to 37 °C).
5. Once cells have lifted off the culture dish, quench the trypsin with complete media and transfer the cells to a sterile 15 ml tube.
6. Centrifuge the harvested cells for 5 minutes at 300 g at room temperature (RT).
7. Remove supernatant and resuspend cells in 1 ml of DMEM or RPMI without additives (depending on cell line).
8. Determine the cell concentration using trypan blue and counting with a hemacytometer.
   (a) For CT2A and GL261, we implant 40,000 cells per mouse.
   (b) For SB28, we implant 1000 cells per mouse.
9. Spin down again and resuspend the cells in a volume of DMEM or RPMI without additives to achieve the desired concentration of 40,000 cells/3 μL for CT2A/GL261 or 1000 cells/3 μL for SB28. We recommend preparing enough cells for twice the number of mice needed. Then, transfer to a 1.7 ml sterile Eppendorf tube and place on ice (Fig. 1, **step 2**).
10. Cells on ice should be briefly vortexed on a low setting or gently flicked after every 3 mice to prevent cell clumping. Cells can be safely used for intracranial implantations up to 2–3 h after placement on ice. Longer times after this will lead to greater cell engraftment variability due to cell death.

3.1.1 Orthotopic Implantation

Note: All animal experiments should be done in accordance with institutional IACUC approved protocols and local standards.

1. Allow mice to acclimate to the environment for at least 1 week prior to tumor implantation. Without acclimation, stressed mice may yield less reproducible results in testing tumor responses to treatment.

2. Prior to implantation, premedicate the mouse with 1 mg/kg of buprenorphine-ER via subcutaneous injection, which serves as prophylactic for post-procedure pain.
3. Anesthetize the mouse with intraperitoneal ketamine/xylazine mixture at a dose of 100 mg/kg and 10 mg/kg, respectively.
4. Apply ophthalmic ointment to the eyes to prevent drying and injury.
5. Shave an 8–10 mm patch of hair overlying the vertex of the mouse's head.
6. Apply 10% betadine solution to the top of the mouse's scalp to disinfect the incision site, taking care to avoid getting the solution into the eyes.
7. Check that the mouse is sufficiently anesthetized by assessing its response to a toe pinch. If the mouse responds to the toe pinch, delay the procedure until the anesthesia can take effect or consider injecting additional anesthesia in small increments.
8. Once the mouse is fully sedated, position it in the Stoelting Mouse Stereotactic Platform (Stoelting Inc.) (Figs. 2 and 3) by hooking the mouse's incisor teeth in the bite bar of the snout restrainer and gently tightening the nose clamp over the snout.
9. Adjust the positioning of the mouse so that the head of the mouse is level and the ear bars line up with the external ear canals bilaterally.
10. Adjust the height of the ear bars as necessary and then advance the ear bars into the caudal portion of the ear canal, gently tightening them such that the mouse's head remains level and secure (Fig. 1, **step 3**).

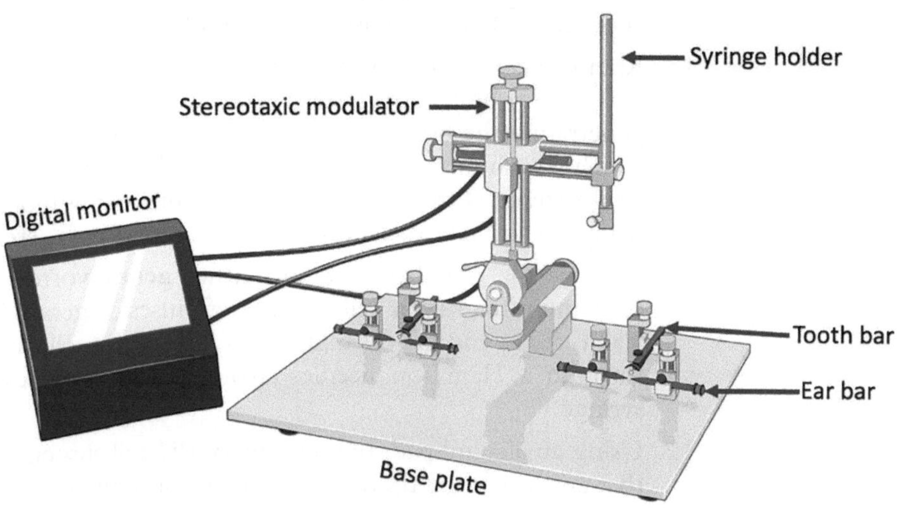

Fig. 2 Stereotaxic instrument setup and its components

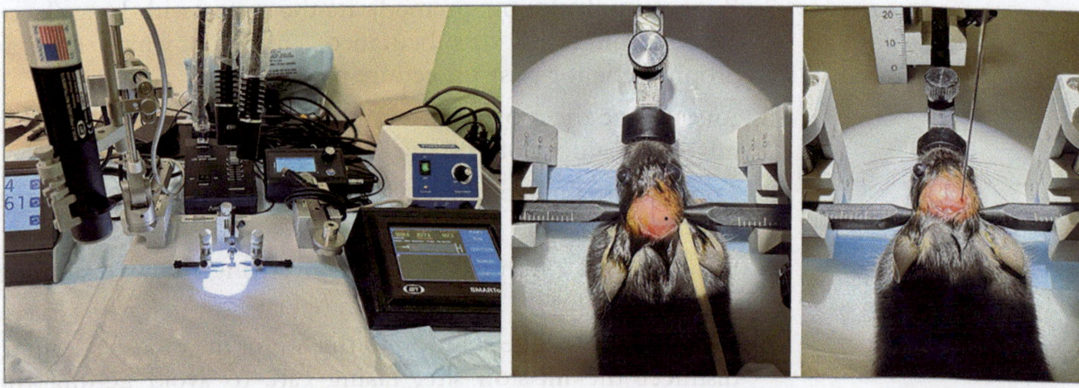

Fig. 3 Stereotactic intracranial tumor implantation. (Left) Stereotaxic instrument setup. (Middle) Mouse in stereotaxic head holder. Black dot marks site of burr hole. (Right) Hamilton needle in burr hole during tumor implantation

11. After positioning is complete, assess for signs of respiratory distress. Loosen and reposition if needed.
12. Using a sterile scalpel, make a 5–8 mm incision longitudinally in the mid-scalp extending from the level of the eyes caudally. Confirm visualization of bregma (the point where the frontal and parietal bones meet) and lambda (where the occipital bone and parietal bones meet) on the skull (Fig. 3).
13. Using the stereotactic platform, place the Hamilton needle tip exactly over the bregma and then zero out the x, y, z coordinates of the digital stereotactic display. For a 6–8-week-old C57BL/6 J mouse, move the tip of the needle to a position 2.5 mm lateral to the bregma in the right cerebral hemisphere. Mark the burr hole site with a fine tip marker (Fig. 3).
14. Clean the 0.45 mm drill bit (Stoelting) attached to the drill (Foredom Microdrill) using an alcohol pad.
15. Rotate the Hamilton needle to the side and use the drill to make a burr hole in the mouse's skull at the marked site. Upon penetrating the bone, you will feel a "give" or a "pop," upon which you should immediately turn off the drill and pull back. Use a sterile cotton swab to gently remove any blood/bone debris to facilitate visualization of the skull (Fig. 1, **step 4**).
16. Remove the cells from ice and either gently vortex the vial containing the tumor cells with brief pulses or gently flick the vial to resuspend the cells. Draw up 5uL of the cell suspension through a Hamilton needle. Avoid any air bubbles in the syringe.
17. Using an alcohol pad, remove any of the cell suspension fluid that appears at the tip of the needle to prevent contamination of the incision site with cancer cells, which may result in the tumor growth in the extracranial space.

18. Rotate the Hamilton syringe back so that it hovers right over the burr hole. Advance the syringe needle into the burr hole maintaining the needle perpendicular to the skull. Once the needle is flush with the skull, zero out the coordinates on the stereotactic digital display, and then slowly advance the tip of the needle until it reaches a depth of 3.0 mm. Leave at that depth for 5 seconds, then retract to 2.5 mm for the implantation.
19. The *x*, *y*, *z* coordinates for stereotactic implantation were chosen to generate a laterally positioned tumor within the cerebral hemisphere while avoiding injury to critical brain structures such as the thalamus and proximity to the ventricles, thus reducing the likelihood of seeding the cerebrospinal fluid with tumor cells.
20. To implant, inject 3 μL of the tumor cell suspension over 60–90 s. Then, leave the needle in place for another 45–60 s prior to slowly withdrawing the needle over 10 s to minimize reflux of tumor cells along the needle tract. After implantation, dry the skull repeatedly with a sterile cotton Q-tip swab to remove any tumor-containing fluid that may have refluxed out of the burr hole.
21. Loosen the ear bars and snout restrainer and remove the mouse from the stereotactic apparatus.
22. Re-approximate the edges of the incision with sterile forceps and close with a horizontal mattress stitch using 4–0 monofilament nylon suture.
23. Place the mouse on a heating pad set to 37 °C until the mouse recovers consciousness. Once the mouse is alert and responsive, transfer it to a clean cage.

3.2 Bioluminescence Imaging for Tumor Confirmation

1. We perform bioluminescence imaging on postimplantation day 6 to confirm tumor engraftment prior to starting treatment (Fig. 1, **step 5**).

   Note: Given the black fur on C57BL/6 J mice, it is important to choose an adequately bright luciferase construct and optimal transduction for visualizing with bioluminescence imaging. In addition, if the luminescence signal remains low/-undetectable, the heads of the mice should be re-shaved prior to imaging.
2. Anesthetize mice using either the ketamine-xylaxine solution or 2% inhaled isoflurane.
3. While the mice are anesthetized, inject them intraperitoneally with 150 mg/kg of D-luciferin. For example, for a 20-gram mouse, give 200 μL of 15 mg/mL D-luciferin solution.
4. Position mice on the imaging stage.

5. Ten minutes after the D-luciferin injection, the mice are imaged with a 1-minute exposure time. This timing may vary and needs to be titrated but should be consistent for all mice on a given day.

   Note: In the case that one mouse has very high BLI signal and other mice has very low or no signal, remove the mouse with a very high BLI and reimage the other mice to avoid any masking of signals.

6. Return the mouse back to the cage once the mouse is alert and responsive.
7. After acquisition of images, use the software program associated with the bioluminescence imager to analyze the tumors. We report our data as total flux (photons/second).

3.3 Intratumoral Mouse-Adapted Zika Virus Treatment

Viral propagation and quantification:

1. Mouse-adapted *Zika virus* (MA-ZIKV) viral aliquots are generated from previously prepared [70, 78, 79] viral stock.
2. Vero cells are grown for at least 2 passages. Seed cells in a T-175 cm tissue culture flask and incubate at 37 °C and 5% CO_2 until cells are 50–60% confluent.
3. Inoculate MA-ZIKV at a multiplicity of infection (MOI) of 0.01 and incubate for 72 hours.
4. Observe cells for cytopathic effect. Collect supernatant 72 hours after infection and centrifuge at 1000G for 10 min to pellet cell debris. Collect supernatant and store at –80 °C.
5. Check viral titer using a plaque assay.
6. Seed Vero cells in a 6-well plate and incubate at 37 °C. Cells should be at least 90% confluent at the time of MA-ZIKV infection.
7. Thaw cryovial containing MA-ZIKV at 4 °C before use.
8. Prepare a serial dilution (10^{-4} to 10^{-8}) of MA-ZIKV in infection media (DMEM+2% fetal bovine serum+1% penicillin/streptomycin).
9. Infect cells with each serial dilution in triplicate and incubate at 37 °C and 5% CO_2 for 2 h.
10. After 2 hours, change media and add 4 ml of 2% methyl cellulose to each well and incubate until plaques are observed (3–5 days).
11. Fix cells with 10% formaldehyde solution for 1 hour, followed by adding enough crystal violet solution to cover the wells. Let it sit for at least 6 hours or overnight. Rinse wells with DI water and air dry to count the plaques.

12. To calculate plaque-forming units per ml (PFU/ml).
13. PFU/ml = (plaque count/well) × (dilution factor) (ml inoculum).

Intratumoral MA-Zika virus injection

1. Mice are serpentine, sorted by BLI intensity in different groups before MA-ZIKV injections.
2. Anesthetize the mouse with an intraperitoneal injection of a ketamine/xylazine mixture at a dose of 100 mg/kg and 10 mg/kg, respectively.
3. Apply ophthalmic ointment to the eyes.
4. Apply betadine solution to the top of the mouse's scalp.
5. Check that the mouse is sufficiently anesthetized by assessing its response to a toe pinch. If the mouse responds to the toe pinch, delay the procedure until the anesthesia can take effect or consider injecting additional anesthesia in small increments.
6. Once fully sedated, position the mouse in the Stoelting Mouse Stereotactic Platform as described above (Fig. 1, **step 6**).
7. Using a sterile scalpel, reopen the prior incision.
8. Position the Hamilton needle so that it hovers right over the burr hole. Advance the syringe needle into the burr hole. Once the needle is flush with the skull, zero out the coordinates on the stereotactic digital display and then slowly advance the tip of the needle until it reaches a depth of 3.0 mm. Leave at that depth for 5 seconds, then retract to 2.5 mm for the injection.
9. Inject 10 μl of the MA-ZIKV suspension over 60 s. Then, leave the needle in place for another 45–60 s prior to slowly withdrawing the needle over 10 seconds (Fig. 1, **step 7**).
10. Loosen the ear bars and snout restrainer, and remove the mouse from the stereotactic apparatus.
11. Re-approximate the edges of the incision with sterile forceps and apply veterinary tissue glue to seal the wound, taking care to avoid glue exposure to the mouse's eyes.

    Note: Veterinary tissue glue can be applied if no further procedures are needed in the next 2–3 weeks. If any other procedures need to be done soon, close the incision using 4–0 monofilament nylon suture.
12. Place the mouse on a heating pad set to 37 °C until the mouse recovers consciousness. Transfer the mouse back to its original cage once it is alert and responsive.

4 Conclusions

We demonstrate that intratumoral MA-ZIKV treatment promotes CD8+ T cell infiltration in the tumor microenvironment. C57BL6/J Mice bearing CT2A glioma were treated with either MA-ZIKV or PBS. After 7 days of MA-ZIKV treatment (or 14 days after tumor implantation), the brain was collected and stained for CD8+ T cells using immunohistochemistry (IHC). IHC images showed decreased tumor size and increased CD8+ T cells in the tumor microenvironment and the neighboring areas after MA-ZIKV treatment compared to PBS control, which has a bigger tumor size and less CD8+ T cells (Fig. 4).

Note: Previous work in our lab demonstrated that intratumoral ZIKV treatment improved survival of mice bearing glioma, in a CD8-dependent manner [69]. Using in situ hybridization, we also observed that MA-ZIKV was largely confined to the tumor [70].

Acknowledgments

MGC was supported by grants R01NS107833 and R01NS117149 from National Institute of Health and by the Alvin J. Siteman Cancer Center Siteman Investment Program through funding from The Foundation for Barnes-Jewish Hospital and the Barnard Trust. YDL was supported by the R38 Stimulating Access to Research in Residency and Dacey Neurosurgical Scholarship grants.

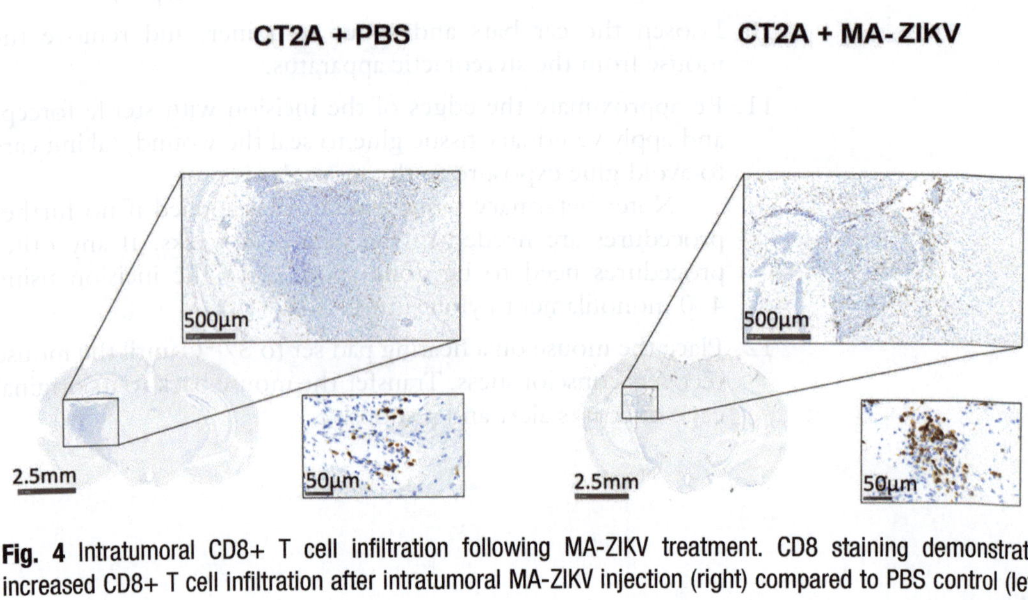

Fig. 4 Intratumoral CD8+ T cell infiltration following MA-ZIKV treatment. CD8 staining demonstrating increased CD8+ T cell infiltration after intratumoral MA-ZIKV injection (right) compared to PBS control (left)

References

1. Kleihues P, Louis DN, Scheithauer BW, Rorke LB, Reifenberger G, Burger PC et al (2002) The WHO classification of tumors of the nervous system. J Neuropathol Exp Neurol 61(3): 215–25; discussion 26–9. https://doi.org/10.1093/jnen/61.3.215
2. Louis DN, Perry A, Reifenberger G, von Deimling A, Figarella-Branger D, Cavenee WK et al (2016) The 2016 World Health Organization classification of tumors of the central nervous system: a summary. Acta Neuropathol 131(6):803–820. https://doi.org/10.1007/s00401-016-1545-1
3. Lenting K, Verhaak R, Ter Laan M, Wesseling P, Leenders W (2017) Glioma: experimental models and reality. Acta Neuropathol 133(2):263–282. https://doi.org/10.1007/s00401-017-1671-4
4. Flynn JR, Wang L, Gillespie DL, Stoddard GJ, Reid JK, Owens J et al (2008) Hypoxia-regulated protein expression, patient characteristics, and preoperative imaging as predictors of survival in adults with glioblastoma multiforme. Cancer 113(5):1032–1042. https://doi.org/10.1002/cncr.23678
5. Hou LC, Veeravagu A, Hsu AR, Tse VC (2006) Recurrent glioblastoma multiforme: a review of natural history and management options. Neurosurg Focus 20(4):E5. https://doi.org/10.3171/foc.2006.20.4.2
6. Nima Derakhshan NAsa-htcu-iffhriwuxhi-e (2019) Cerebellar glioblastoma multiforme in an adult patient with neurofibromatosis type 1: an extremely rare report with review of literature. Egypt J Neurol Psych Neurosurg 55
7. Rees JH, Smirniotopoulos JG, Jones RV, Wong K (1996) Glioblastoma multiforme: radiologic-pathologic correlation. Radiographics 16(6):1413–38; quiz 62–3. https://doi.org/10.1148/radiographics.16.6.8946545
8. Sponagel J, Jones JK, Frankfater C, Zhang S, Tung O, Cho K et al (2022) Sex differences in brain tumor glutamine metabolism reveal sex-specific vulnerabilities to treatment. Med 3(11):792–811.e12. https://doi.org/10.1016/j.medj.2022.08.005
9. Lee J, Nicosia M, Hong ES, Silver DJ, Li C, Bayik D et al (2023) Sex-biased T-cell exhaustion drives differential immune responses in glioblastoma. Cancer Discov 13(9): 2090–2105. https://doi.org/10.1158/2159-8290.cd-22-0869
10. Pombo Antunes AR, Scheyltjens I, Duerinck J, Neyns B, Movahedi K, Van Ginderachter JA (2020) Understanding the glioblastoma immune microenvironment as basis for the development of new immunotherapeutic strategies. elife 9. https://doi.org/10.7554/eLife.52176
11. Chen Z, Hambardzumyan D (2018) Immune microenvironment in glioblastoma subtypes. Front Immunol 9:1004. https://doi.org/10.3389/fimmu.2018.01004
12. Lim M, Weller M, Idbaih A, Steinbach J, Finocchiaro G, Raval RR et al (2022) Phase III trial of chemoradiotherapy with temozolomide plus nivolumab or placebo for newly diagnosed glioblastoma with methylated MGMT promoter. Neuro-Oncology 24(11): 1935–1949. https://doi.org/10.1093/neuonc/noac116
13. Reardon DA, Brandes AA, Omuro A, Mulholland P, Lim M, Wick A et al (2020) Effect of Nivolumab vs bevacizumab in patients with recurrent glioblastoma: the CheckMate 143 phase 3 randomized clinical trial. JAMA Oncol 6(7):1003–1010. https://doi.org/10.1001/jamaoncol.2020.1024
14. Gimple RC, Bhargava S, Dixit D, Rich JN (2019) Glioblastoma stem cells: lessons from the tumor hierarchy in a lethal cancer. Genes Dev 33(11–12):591–609. https://doi.org/10.1101/gad.324301.119
15. Galdieri L, Jash A, Malkova O, Mao DD, DeSouza P, Chu YE et al (2021) Defining phenotypic and functional heterogeneity of glioblastoma stem cells by mass cytometry. JCI. Insight 6(4). https://doi.org/10.1172/jci.insight.128456
16. Xie XP, Laks DR, Sun D, Ganbold M, Wang Z, Pedraza AM et al (2022) Quiescent human glioblastoma cancer stem cells drive tumor initiation, expansion, and recurrence following chemotherapy. Dev Cell 57(1):32–46.e8. https://doi.org/10.1016/j.devcel.2021.12.007
17. Lee J, Jo DH, Kim JH, Cho CS, Han JE, Kim Y et al (2019) Development of a patient-derived xenograft model of glioblastoma via intravitreal injection in mice. Exp Mol Med 51(4):1–9. https://doi.org/10.1038/s12276-019-0241-3
18. Jin KT, Du WL, Lan HR, Liu YY, Mao CS, Du JL et al (2021) Development of humanized mouse with patient-derived xenografts for cancer immunotherapy studies: a comprehensive review. Cancer Sci 112(7):2592–2606. https://doi.org/10.1111/cas.14934
19. Okada S, Vaeteewoottacharn K, Kariya R (2019) Application of highly immunocompromised mice for the establishment of patient-derived xenograft (PDX) models. Cells 8(8). https://doi.org/10.3390/cells8080889

20. Zeng W, Tang Z, Li Y, Yin G, Liu Z, Gao J et al (2020) Patient-derived xenografts of different grade gliomas retain the heterogeneous histological and genetic features of human gliomas. Cancer Cell Int 20:1. https://doi.org/10.1186/s12935-019-1086-5
21. Zeng M, Ruan Z, Tang J, Liu M, Hu C, Fan P et al (2023) Generation, evolution, interfering factors, applications, and challenges of patient-derived xenograft models in immunodeficient mice. Cancer Cell Int 23 1:120. https://doi.org/10.1186/s12935-023-02953-3
22. Yang QE (2021) Human cancer xenografts in immunocompromised mice provide an advanced genuine tumor model for research and drug development-a revisit of murine models for human cancers. Biochim Biophys Acta Gen Subj 1865(8):129929. https://doi.org/10.1016/j.bbagen.2021.129929
23. Srivastava R, Labani-Motlagh A, Chen A, Bohorquez JA, Qin B, Dodda M et al (2023) Development of a human glioblastoma model using humanized DRAG mice for immunotherapy. Antib Ther 6(4):253–264. https://doi.org/10.1093/abt/tbad021
24. Hatcher A, Yu K, Meyer J, Aiba I, Deneen B, Noebels JL (2020) Pathogenesis of peritumoral hyperexcitability in an immunocompetent CRISPR-based glioblastoma model. J Clin Invest 130(5):2286–2300. https://doi.org/10.1172/jci133316
25. Kijima N, Kanemura Y (2017) Mouse models of glioblastoma. In: De Vleeschouwer S (ed) Glioblastoma. Codon Publications Copyright: The Authors, Brisbane (AU)
26. Olson B, Li Y, Lin Y, Liu ET, Patnaik A (2018) Mouse models for cancer immunotherapy research. Cancer Discov 8(11):1358–1365. https://doi.org/10.1158/2159-8290.cd-18-0044
27. Ausman JI, Shapiro WR, Rall DP (1970) Studies on the chemotherapy of experimental brain tumors: development of an experimental model. Cancer Res 30(9):2394–2400
28. Genoud V, Marinari E, Nikolaev SI, Castle JC, Bukur V, Dietrich PY et al (2018) Responsiveness to anti-PD-1 and anti-CTLA-4 immune checkpoint blockade in SB28 and GL261 mouse glioma models. Onco Targets Ther 7 12:e1501137. https://doi.org/10.1080/2162402x.2018.1501137
29. Touat M, Li YY, Boynton AN, Spurr LF, Iorgulescu JB, Bohrson CL et al (2020) Mechanisms and therapeutic implications of hypermutation in gliomas. Nature 580(7804):517–523. https://doi.org/10.1038/s41586-020-2209-9
30. Qi Y, Liu B, Sun Q, Xiong X, Chen Q (2020) Immune checkpoint targeted therapy in glioma: status and hopes. Front Immunol 11:578877. https://doi.org/10.3389/fimmu.2020.578877
31. Johanns TM, Ward JP, Miller CA, Wilson C, Kobayashi DK, Bender D et al (2016) Endogenous Neoantigen-specific CD8 T cells identified in two glioblastoma models using a cancer Immunogenomics approach. Cancer Immunol Res 4(12):1007–1015. https://doi.org/10.1158/2326-6066.cir-16-0156
32. Le DT, Uram JN, Wang H, Bartlett BR, Kemberling H, Eyring AD et al (2015) PD-1 blockade in tumors with mismatch-repair deficiency. N Engl J Med 372(26):2509–2520. https://doi.org/10.1056/NEJMoa1500596
33. Llosa NJ, Cruise M, Tam A, Wicks EC, Hechenbleikner EM, Taube JM et al (2015) The vigorous immune microenvironment of microsatellite instable colon cancer is balanced by multiple counter-inhibitory checkpoints. Cancer Discov 5(1):43–51. https://doi.org/10.1158/2159-8290.cd-14-0863
34. Iorgulescu JB, Gokhale PC, Speranza MC, Eschle BK, Poitras MJ, Wilkens MK et al (2021) Concurrent dexamethasone limits the clinical benefit of immune checkpoint blockade in glioblastoma. Clin Cancer Res 27(1):276–287. https://doi.org/10.1158/1078-0432.ccr-20-2291
35. Xue W, Zhang J, Tong H, Xie T, Chen X, Zhou B et al (2020) Effects of BMPER, CXCL10, and HOXA9 on neovascularization during early-growth stage of primary high-grade glioma and their corresponding MRI biomarkers. Front Oncol 10:711. https://doi.org/10.3389/fonc.2020.00711
36. Sharma P, Allison JP (2015) The future of immune checkpoint therapy. Science 348(6230):56–61. https://doi.org/10.1126/science.aaa8172
37. Wiesner SM, Decker SA, Larson JD, Ericson K, Forster C, Gallardo JL et al (2009) De novo induction of genetically engineered brain tumors in mice using plasmid DNA. Cancer Res 69(2):431–439. https://doi.org/10.1158/0008-5472.can-08-1800
38. Kosaka A, Ohkuri T, Okada H (2014) Combination of an agonistic anti-CD40 monoclonal antibody and the COX-2 inhibitor celecoxib induces anti-glioma effects by promotion of type-1 immunity in myeloid cells and T-cells. Cancer Immunol Immunother 63(8):847–857. https://doi.org/10.1007/s00262-014-1561-8
39. Kohanbash G, Carrera DA, Shrivastav S, Ahn BJ, Jahan N, Mazor T et al (2017) Isocitrate dehydrogenase mutations suppress STAT1 and CD8+ T cell accumulation in gliomas. J Clin Invest 127(4):1425–1437. https://doi.org/10.1172/jci90644

40. Leach DR, Krummel MF, Allison JP (1996) Enhancement of antitumor immunity by CTLA-4 blockade. Science 271(5256): 1734–1736. https://doi.org/10.1126/science.271.5256.1734
41. Wainwright DA, Chang AL, Dey M, Balyasnikova IV, Kim CK, Tobias A et al (2014) Durable therapeutic efficacy utilizing combinatorial blockade against IDO, CTLA-4, and PD-L1 in mice with brain tumors. Clin Cancer Res 20(20):5290–5301. https://doi.org/10.1158/1078-0432.ccr-14-0514
42. Reardon DA, Gokhale PC, Klein SR, Ligon KL, Rodig SJ, Ramkissoon SH et al (2016) Glioblastoma eradication following immune checkpoint blockade in an Orthotopic, Immunocompetent Model Cancer. Immunol Res 4(2):124–135. https://doi.org/10.1158/2326-6066.cir-15-0151
43. Tykocki T, Eltayeb M (2018) Ten-year survival in glioblastoma. A systematic review. J Clin Neurosci 54:7–13. https://doi.org/10.1016/j.jocn.2018.05.002
44. Oh T, Fakurnejad S, Sayegh ET, Clark AJ, Ivan ME, Sun MZ et al (2014) Immunocompetent murine models for the study of glioblastoma immunotherapy. J Transl Med 12:107. https://doi.org/10.1186/1479-5876-12-107
45. Seyfried TN, el-Abbadi M, Roy ML (1992) Ganglioside distribution in murine neural tumors. Mol Chem Neuropathol 17(2): 147–167. https://doi.org/10.1007/bf03159989
46. Khaled YS, Ammori BJ, Elkord E (2013) Myeloid-derived suppressor cells in cancer: recent progress and prospects. Immunol Cell Biol 91(8):493–502. https://doi.org/10.1038/icb.2013.29
47. Martínez-Murillo R, Martínez A (2007) Standardization of an orthotopic mouse brain tumor model following transplantation of CT-2A astrocytoma cells. Histol Histopathol 22(12):1309–1326. https://doi.org/10.14670/hh-22.1309
48. Belmans J, Van Woensel M, Creyns B, Dejaegher J, Bullens DM, Van Gool SW (2017) Immunotherapy with subcutaneous immunogenic autologous tumor lysate increases murine glioblastoma survival. Sci Rep 7 1:13902. https://doi.org/10.1038/s41598-017-12584-0
49. Binello E, Qadeer ZA, Kothari HP, Emdad L, Germano IM (2012) Stemness of the CT-2A immunocompetent mouse brain tumor model: characterization in vitro. J Cancer 3:166–174. https://doi.org/10.7150/jca.4149
50. Zeppernick F, Ahmadi R, Campos B, Dictus C, Helmke BM, Becker N et al (2008) Stem cell marker CD133 affects clinical outcome in glioma patients. Clin Cancer Res 14(1): 123–129. https://doi.org/10.1158/1078-0432.ccr-07-0932
51. Riva M, Wouters R, Weerasekera A, Belderbos S, Nittner D, Thal DR et al (2019) CT-2A neurospheres-derived high-grade glioma in mice: a new model to address tumor stem cells and immunosuppression. Biol Open 8(9). https://doi.org/10.1242/bio.044552
52. Liu CJ, Schaettler M, Blaha DT, Bowman-Kirigin JA, Kobayashi DK, Livingstone AJ et al (2020) Treatment of an aggressive orthotopic murine glioblastoma model with combination checkpoint blockade and a multivalent neoantigen vaccine. Neuro-Oncology 22(9): 1276–1288. https://doi.org/10.1093/neuonc/noaa050
53. Nakashima H, Alayo QA, Penaloza-MacMaster P, Freeman GJ, Kuchroo VK, Reardon DA et al (2018) Modeling tumor immunity of mouse glioblastoma by exhausted CD8 (+) T cells. Sci Rep 8 1:208. https://doi.org/10.1038/s41598-017-18540-2
54. Romani M, Pistillo MP, Carosio R, Morabito A, Banelli B (2018) Immune checkpoints and innovative therapies in glioblastoma. Front Oncol 8:464. https://doi.org/10.3389/fonc.2018.00464
55. Barnard Z, Wakimoto H, Zaupa C, Patel AP, Klehm J, Martuza RL et al (2012) Expression of FMS-like tyrosine kinase 3 ligand by oncolytic herpes simplex virus type I prolongs survival in mice bearing established syngeneic intracranial malignant glioma. Neurosurgery 71(3):741–748; discussion 8;. https://doi.org/10.1227/NEU.0b013e318260fd73
56. Dutter A, Weissberg M, Wrba H (1982) Induction of a slowly growing transplantable adenocarcinoma metastasizing into the lungs. Exp Pathol 22(3):179–180. https://doi.org/10.1016/s0232-1513(82)80006-6
57. Kaufman HL, Kohlhapp FJ, Zloza A (2015) Oncolytic viruses: a new class of immunotherapy drugs. Nat Rev Drug Discov 14(9): 642–662. https://doi.org/10.1038/nrd4663
58. Haseley A, Alvarez-Breckenridge C, Chaudhury AR, Kaur B (2009) Advances in oncolytic virus therapy for glioma. Recent Pat CNS Drug Discov 4(1):1–13. https://doi.org/10.2174/157488909787002573
59. Chadene Z (2021) Tremaglio a JNBb, Rachel Fearns c. Chapter 2 – Genetic instability of RNA viruses. Academic
60. Krasnykh VN, Mikheeva GV, Douglas JT, Curiel DT (1996) Generation of recombinant adenovirus vectors with modified fibers for altering viral tropism. J Virol 70(10):6839–6846. https://doi.org/10.1128/jvi.70.10.6839-6846.1996

61. Lamfers ML, Grill J, Dirven CM, Van Beusechem VW, Geoerger B, Van Den Berg J et al (2002) Potential of the conditionally replicative adenovirus Ad5-Delta24RGD in the treatment of malignant gliomas and its enhanced effect with radiotherapy. Cancer Res 62(20): 5736–5742
62. Chiocca EA, Nakashima H, Kasai K, Fernandez SA, Oglesbee M (2020) Preclinical toxicology of rQNestin34.5v.2: an oncolytic herpes virus with transcriptional regulation of the ICP34.5 Neurovirulence gene. Mol Ther Methods Clin Dev 17:871–893. https://doi.org/10.1016/j.omtm.2020.03.028
63. Engeland CE, Ungerechts G (2021) Measles virus as an oncolytic immunotherapy. Cancers (Basel) 13(3). https://doi.org/10.3390/cancers13030544
64. Ganly I, Kirn D, Eckhardt G, Rodriguez GI, Soutar DS, Otto R et al (2000) A phase I study of Onyx-015, an E1B attenuated adenovirus, administered intratumorally to patients with recurrent head and neck cancer. Clin Cancer Res 6(3):798–806
65. Park SH, Breitbach CJ, Lee J, Park JO, Lim HY, Kang WK et al (2015) Phase 1b trial of biweekly intravenous Pexa-Vec (JX-594), an oncolytic and immunotherapeutic vaccinia virus in colorectal cancer. Mol Ther 23(9): 1532–1540. https://doi.org/10.1038/mt.2015.109
66. Nassiri F, Patil V, Yefet LS, Singh O, Liu J, Dang RMA et al (2023) Oncolytic DNX-2401 virotherapy plus pembrolizumab in recurrent glioblastoma: a phase 1/2 trial. Nat Med 29(6):1370–1378. https://doi.org/10.1038/s41591-023-02347-y
67. Lang FF, Conrad C, Gomez-Manzano C, Yung WKA, Sawaya R, Weinberg JS et al (2018) Phase I study of DNX-2401 (Delta-24-RGD) oncolytic adenovirus: replication and immunotherapeutic effects in recurrent malignant glioma. J Clin Oncol 36(14):1419–1427. https://doi.org/10.1200/jco.2017.75.8219
68. Rehman H, Silk AW, Kane MP, Kaufman HL (2016) Into the clinic: Talimogene laherparepvec (T-VEC), a first-in-class intratumoral oncolytic viral therapy. J Immunother Cancer 4:53. https://doi.org/10.1186/s40425-016-0158-5
69. Nair S, Mazzoccoli L, Jash A, Govero J, Bais SS, Hu T et al (2021) Zika virus oncolytic activity requires CD8+ T cells and is boosted by immune checkpoint blockade. JCI. Insight 6(1). https://doi.org/10.1172/jci.insight.144619
70. Zhu Z, Gorman MJ, McKenzie LD, Chai JN, Hubert CG, Prager BC et al (2017) Zika virus has oncolytic activity against glioblastoma stem cells. J Exp Med 214(10):2843–2857. https://doi.org/10.1084/jem.20171093
71. Shen Z, Liu X, Fan G, Na J, Liu Q, Lin F et al (2023) Improving the therapeutic efficacy of oncolytic viruses for cancer: targeting macrophages. J Transl Med 21 1:842. https://doi.org/10.1186/s12967-023-04709-z
72. Kooti W, Esmaeili Gouvarchin Ghaleh H, Farzanehpour M, Dorostkar R, Jalali Kondori B, Bolandian M (2021) Oncolytic viruses and cancer, do you know the Main mechanism? Front. Oncologia 11:761015. https://doi.org/10.3389/fonc.2021.761015
73. Mardi A, Shirokova AV, Mohammed RN, Keshavarz A, Zekiy AO, Thangavelu L et al (2022) Biological causes of immunogenic cancer cell death (ICD) and anti-tumor therapy; combination of oncolytic virus-based immunotherapy and CAR T-cell therapy for ICD induction. Cancer Cell Int 22 1:168. https://doi.org/10.1186/s12935-022-02585-z
74. Ma J, Ramachandran M, Jin C, Quijano-Rubio C, Martikainen M, Yu D et al (2020) Characterization of virus-mediated immunogenic cancer cell death and the consequences for oncolytic virus-based immunotherapy of cancer. Cell Death Dis 11(1):48. https://doi.org/10.1038/s41419-020-2236-3
75. Bin Umair M, Akusa FN, Kashif H, Seerat EF, Butt F, Azhar M et al (2022) Viruses as tools in gene therapy, vaccine development, and cancer treatment. Arch Virol 167(6):1387–1404. https://doi.org/10.1007/s00705-022-05432-8
76. Ma R, Li Z, Chiocca EA, Caligiuri MA, Yu J (2023) The emerging field of oncolytic virus-based cancer immunotherapy. Trends Cancer 9(2):122–139. https://doi.org/10.1016/j.trecan.2022.10.003
77. Shalhout SZ, Miller DM, Emerick KS, Kaufman HL (2023) Therapy with oncolytic viruses: progress and challenges. In: Nat Rev Clin Oncol. England: © 2023. Springer, pp 160–177
78. Xie X, Kum DB, Xia H, Luo H, Shan C, Zou J et al (2018) A single-dose live-attenuated Zika virus vaccine with controlled infection rounds that protects against vertical transmission. Cell Host Microbe 24(4):487–99.e5. https://doi.org/10.1016/j.chom.2018.09.008
79. Gorman MJ, Caine EA, Zaitsev K, Begley MC, Weger-Lucarelli J, Uccellini MB et al (2018) An immunocompetent mouse model of Zika virus infection. Cell Host Microbe 23(5): 672–85.e6. https://doi.org/10.1016/j.chom.2018.04.003

Chapter 4

A Method for Circulating Tumor DNA Sequencing Using Cerebrospinal Fluid in Pediatric Central Nervous System Tumors

Eirini Christodoulou, Dejerianne Ostrow, David Ruble, Yuxia Zhan, Dong Xu, Venkata Yellapantula, and Katrina O'Halloran

Abstract

Central nervous system tumors are a major cause of morbidity and mortality in children. As tumor sequencing has become available, the World Health Organization diagnostic criteria have been refined to include genomic tumor characteristics. Additionally, therapeutic agents targeting specific molecular aberrations are increasingly being developed. Currently, the standard of care for tumor diagnosis includes invasive acquisition of tumor material for histopathology and molecular sequencing. Circulating tumor DNA has been shown to be detectable within the cerebrospinal fluid (CSF) of patients with brain and spinal cord tumors, representing a potential liquid biopsy. In this chapter, we describe our combined platform of Low-Pass Whole Genome Sequencing and targeted sequencing for CSF sample acquisition, cell-free DNA extraction, and processing with success in the detection of DNA copy number aberrations, sequence variants, and fusions, indicative of circulating tumor DNA presence. This technology is being implemented for clinical use at the Center for Personalized Medicine of Children's Hospital Los Angeles.

Key words Cerebrospinal fluid, Liquid biopsy, Brain tumor, Spinal cord tumor, Pediatric neuro-oncology

1 Introduction

Central nervous system (CNS) tumors in children continue to be a major cause of morbidity and mortality. In children and adolescents aged 0–19 years, the incidence of CNS tumors between 2015 and 2019 was reported at 6.20 per 100,000 population [1]. Historically, diagnosis has relied on histopathologic evaluation, which requires invasive tissue acquisition. As sequencing technology has become available, the World Health Organization (WHO) classification of

Michael Goldstein (ed.), *Translational Research Methods for Pediatric Brain Tumors*, Neuromethods, vol. 226,
https://doi.org/10.1007/978-1-0716-4754-7_4,

CNS tumors has been revised to include molecular tumor features [2–4]. This also includes new tumor types, which have been introduced on the basis of molecular characterizations, for example, diffuse midline glioma, H3 K27M-mutant—an unresectable and incurable childhood tumor [3]. As molecular features become more clearly defined, novel agents with specific molecular targets are increasingly being developed [5].

Liquid biopsy is an emerging field with the potential to transform patient care. As noted above, the cytogenetic and molecular genetic characterization of CNS tumors is used clinically as an aid for diagnosis, determining prognosis, and guiding treatment. The main source and best-characterized liquid biopsy biofluid is blood plasma, and various studies have shown the clinical utility of plasma, mainly in the molecular characterization of adult tumor types [6]. In contrast, pediatric cancer is rare and highly heterogeneous; therefore, there is a need to determine the features that drive pediatric oncogenesis [7]. Circulating tumor DNA (ctDNA) from cerebrospinal fluid (CSF) liquid biopsy in patients with CNS tumors has been shown to be detectable in the form of both copy number aberrations (CNA) and sequence variants (SV) [8–10]. CSF has been shown to be a more enriched source of ctDNA, and therefore to have greater clinical utility in CNS tumor types when compared to blood plasma—likely due to the impact of the blood-brain barrier [11, 12].

Potential applications of CSF liquid biopsy include in minimally invasive diagnosis, as in the case of inoperable tumors, which may otherwise not have an indication for surgery. Liquid biopsies may also be utilized in the assessment of response to therapies, as has been reported with novel therapies, and also in surveillance for potential molecular tumor evolution [13]. Furthermore, the predictive value of liquid biopsy with regard to risk for relapse has also been demonstrated in medulloblastoma, where a positive liquid biopsy at the end of therapy was shown to place a child at a significantly higher risk for relapse [14].

A combined Low-Pass Whole Genome Sequencing (LP-WGS) and targeted sequencing platform, as described below, has been established for use at the Children's Hospital Los Angeles Center for Personalized Medicine. This platform utilizes CSF (average 3–5 mL) from pediatric patients with CNS tumors. A single sample undergoes cell-free DNA (cfDNA) extraction and library preparation, which is then split into two aliquots for CNA and SV detection in parallel; CNA detection is now validated with return of results to the ordering clinician, and validation of SV detection is ongoing [15, 16].

2 Materials

2.1 QuantiFluor® ONE dsDNA System

1. Vendor: Promega.
2. p/n: E4871,
3. Description: The QuantiFluor® ONE dsDNA System is a very sensitive method that allows dsDNA quantification for low-concentration samples, such as CSF-derived cfDNA. The sensitivity range of the QuantiFluor® ONE Dye is 0.2–400 ng/μl (1 μl cfDNA sample is diluted in a 200 μl assay).

2.2 Tape Station 2200

1. Vendor: Agilent.
2. p/n: G2965AA,
3. Description: The Agilent 2200 TapeStation system is a tape-based platform for electrophoresis, allowing confirmation of the abundance and molecular size range of a cf-DNA extract using the cf-DNA Screen Tape analysis. The expected molecular size of a CSF-derived cf-DNA extract is ~200 bp. The Tape Station 2200 system is also used to quantify the cfDNA-built sequencing libraries as a Quality Control step prior to sequencing. For the purposes of this protocol, the High Sensitivity Screen Tape is implemented where 2 μl of library is tested with 2 μl sample buffer. The expected molecular library range for a Low-Pass Whole Genome Sequencing sample is ~300 bp.

2.3 NextSeq 500 and HiSeq 4000

1. Vendor: Illumina.
2. p/n: SY-415-1001.
3. Description: The NextSeq Series is a high-throughput sequencing platform, generating up to 800 million paired-end reads with a run time of 1.5 days. The HiSeq instrument allows for higher-throughput sequencing, producing 5 billion paired-end reads per flow cell with a run time of 3.5 days. Sodium hydroxide (NaOH) is a critical reagent during sequencing loading that denatures double-stranded libraries to single-stranded libraries allowing binding to sequencing flow-cells by hybridization. The cf-DNA built libraries are pooled and sequenced at 2 × 100 bp.

2.4 GeneAmp PCR System 9700 or Comparable PCR Thermal Cycler

1. Vendor: Thermo Scientific.
2. p/n: N805–0200,
3. Description: The PCR Thermal Cycler is used for the incubation steps during the library preparation protocol of the CSF-derived cfDNA molecules, including end repair, ligation, and Polymerase Chain Reaction (PCR) amplification of cf-DNA built libraries before and after hybridization to targeted sequencing probes.

3 Methods

3.1 CSF Sample Processing

1. Cerebrospinal fluid (CSF) is collected into sterile specimen tubes (3–5 mL) from any CSF-liberating procedure, for example, lumbar puncture, and is immediately brought to the lab for processing.
2. CSF is immediately centrifuged at 3000× *g* for 10 min at 4 °C to separate the supernatant and CSF pellet.
3. CSF supernatant and pellet are immediately frozen at −80 °C.

3.2 cfDNA Extraction and Quantification

1. Various commercial kits are available for cfDNA extraction, and we recommend both MagMax Cell-Free Total Nucleic Acid Isolation Kit (ThermoFisher, Waltham, MA), a bead-based approach, as well as QIAamp Circulating Nucleic Acid Kit (Qiagen), a column-based approach.
2. The abundance and quality of cfDNA in the extracted samples are assessed using Cell-free DNA ScreenTape analysis (Cat.# 5067-5631; 5067-5630, Agilent Santa Clara, CA) and QuantiFluor® ONE dsDNA System (Promega Madison, WI).

3.3 Library Preparation

1. Libraries are constructed with the xGen cfDNA and FFPE DNA Library Prep Kit (Integrated DNA Technologies, Coralville, IA) according to the manufacturer's instructions using 5 ng of cfDNA input. Library construction includes the addition of Unique Molecular Identifier indexes (UMIs).
2. Libraries are split into two aliquots, one for Low-Pass Whole Genome Sequencing (LP-WGS), the other for hybrid gene capture (targeted sequencing) using the custom panel.
3. The concentration and quality of cfDNA-built libraries are assessed using High-Sensitivity D1000 ScreenTape analysis (Cat.# 5067-5585; 5067-5592, Agilent Santa Clara, CA). In the case of large fragment DNA presence, the libraries undergo a double-sided bead purification as an additional cleanup step for more accurate library concentration count.

3.4 Hybrid Gene Capture

1. 187 ng from each of 8 cfDNA-built libraries is combined to create an equimolar pool.
2. The pool is hybridized to a custom hybrid gene capture panel manufactured by Twist Bioscience (San Francisco, CA). Probes were designed to cover specific genomic regions frequently mutated in gliomas (Table 1). The Twist Exome FastHyb Protocol was used, following the manufacturer's protocol for hybrid gene capture.

Table 1
Genes in a custom hybrid gene capture panel

Gene	Hotspot	Whole gene	Fusion
ACVR1	X		
ATRX		X	
BRAF	X		X
CDKN2A		X	
CDKN2B		X	
EGFR	X		
FGFR1	X		
H3C2	X		
H3-3A	X		
IDH1	X		
IDH2	X		
PDGFRA	X		
PDGFRB	X		
PIK3CA		X	
PIK3R1		X	
TERT	X		
TP53		X	

3.5 Sequencing

1. Double-stranded sequencing libraries are combined to create an equimolar pool and denatured to single-stranded libraries prior to loading on the sequencing instrument using NaOH.
2. The libraries are loaded and sequenced at 2 × 100 bp on an Illumina NextSeq 500, HiSeq 4000, or the NovaSeq series platforms (San Diego, CA).
3. For Low-Pass Whole Genome Sequencing (LP-WGS) to identify Copy Number Alterations (CNAs), the target sequencing depth is 4× (approximately 50 M read pairs per sample).
4. For the hybrid gene capture panel, the target sequencing depth is 1000× (approximately 50 M read pairs per sample).

A graphic representation of the CSF liquid biopsy workflow is shown in Fig. 1.

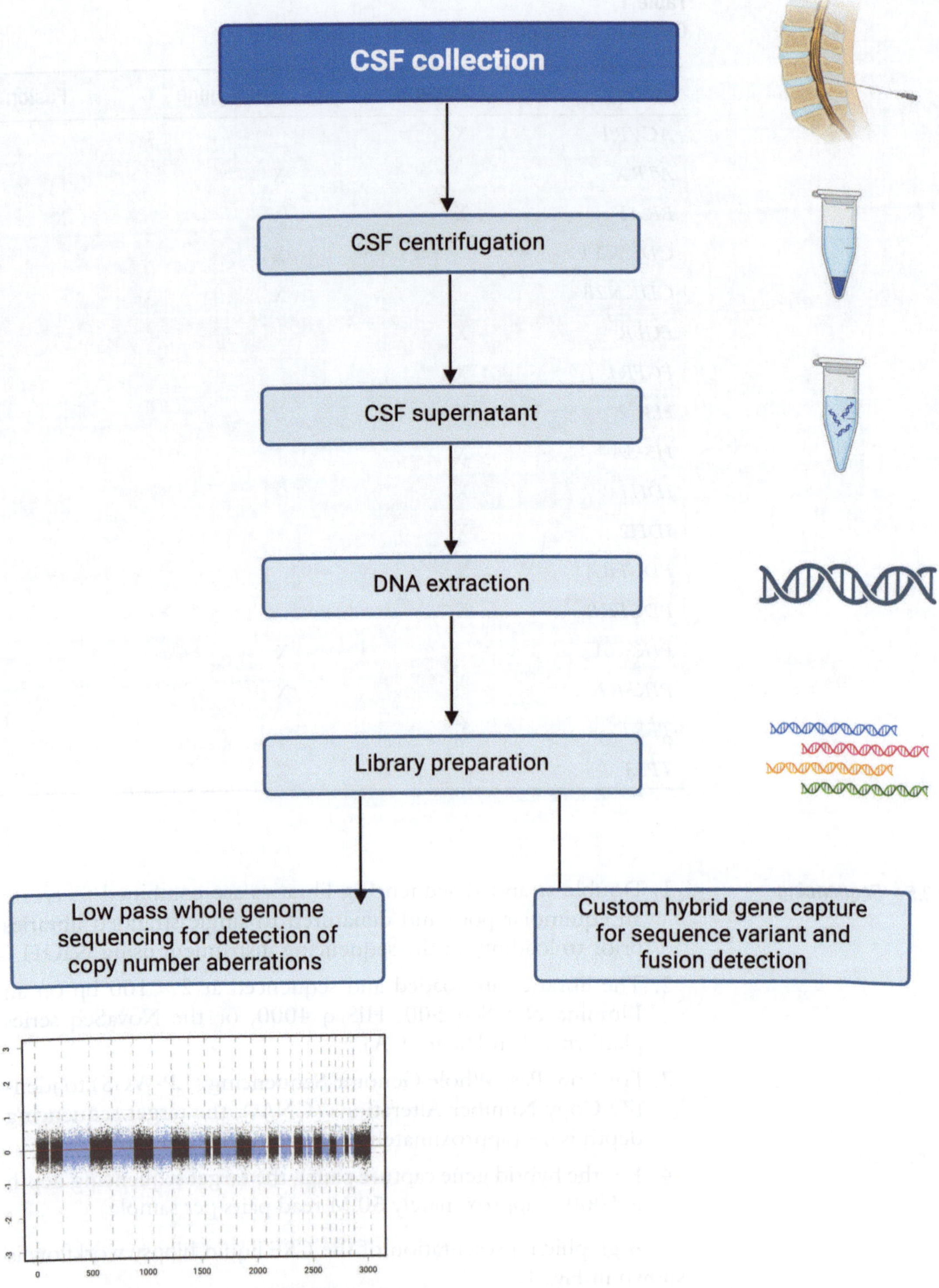

Fig. 1 Representation of the CSF liquid biopsy workflow

3.6 Bioinformatic Analysis

1. Analysis is performed using Illumina's Dragen Tool (v4.1.0). Reads are aligned to the 1000 genomes phase 2 reference genome (hs37d5), which includes build GRCh37 and decoy sequences:

   (ftp://ftptrace.ncbi.nih.gov/1000genomes/ftp/technical/reference/phase2_reference_assembly_sequence/hs37d5.fa.gz)
2. LP-WGS reads are binned into 500 kb regions across the entire genome, and the ichorCNA algorithm is utilized to obtain the segmented calls [17]. Tumor fraction is estimated for all samples using the ichorCNA variant calling pipeline. Briefly, the algorithm uses a Hidden Markov Model (HMM) to predict copy number segments and to estimate the circulating tumor content from an LP-WGS library. An estimated tumor fraction of at least 7% is recommended for CNA positivity in a CSF specimen. To facilitate circulating tumor DNA detection, we recommend performing insert-size selection for sequencing fragments between 90 and 150 bp.
3. For hybrid gene capture, the Dragen somatic pipeline for SNV/Indel calling in tumor samples is implemented [18]. The resultant VCF files are uploaded into Alissa Interpret (Agilent Technologies, Santa Clara, CA) for downstream interpretation. A variant allele fraction (VAF) of 1% is used as a limit-of-detection threshold. The presence of sequence variants is confirmed using the Integrative Genomics Viewer (IGV) [19].
4. For detection of *KIAA1549::BRAF* tandem duplication in low-grade gliomas, UMI's are trimmed using the Dragen 3.7.3 aligner, and the reads are realigned to build 37 of the human reference genome. This increases sensitivity for fusion detection. Illumina Manta (version 1.6.0) is used in the "targeted" mode (https://github.com/Illumina/manta) to detect structural variants.
5. All statistical analysis is carried out using *R* version 4.0.4.

   An example of copy number profiling from multiple primary tumors and corresponding liquid biopsies is shown in Fig. 2.

4 Notes

1. CSF may be collected from any CSF-liberating procedure, for example, lumbar puncture, Ommaya reservoir tap, external ventricular drain, etc.
2. CSF may be collected into sterile CSF specimen containers. Streck tubes may facilitate the stability of sample.

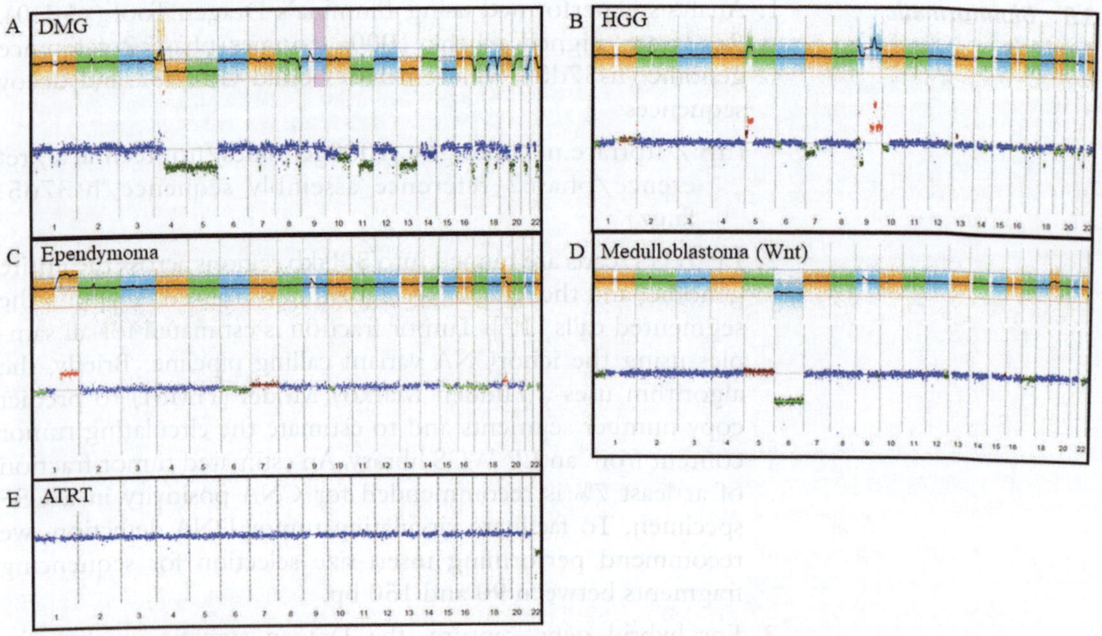

Fig. 2 Representative primary tumor (top) and liquid biopsy iChor (bottom) copy number profiling from variable pediatric central nervous system tumor histopathological diagnoses. The log2 ratio is plotted (y-axis) against chromosome numbers 1–22 (x-axis). Red indicates a copy number gain, green a copy number loss, and blue a neutral copy number. (**a**) Diffuse midline glioma with PDGFRA amplification and various additional whole and partial chromosome losses. (**b**) High-grade glioma with 5p gain, complex 9 gains and losses, and various additional changes. (**c**) Ependymoma with 1q gain, a poor prognostic marker. (**d**) Medulloblastoma with monosomy 6, consistent with the Wnt subgroup. E AT/RT with chromosome 22 deletion, harboring the *SMARCB1* gene

3. We recommend minimizing the time from sample acquisition to centrifugation and freezing to not exceed 1 h.
4. We recommend 5 ng cfDNA input for sequencing library construction; however, the protocol may still be successfully completed with as low as 2 ng cfDNA.
5. Liquid biopsy findings may be compared to primary tumor molecular analyses, when available.

Conflict of Interest

The authors have no conflicts of interest to declare that are relevant to the content of this chapter.

References

1. Ostrom QT et al (2022) CBTRUS statistical report: primary brain and other central nervous system tumors diagnosed in the United States in 2015-2019. Neuro-Oncology 24(Suppl 5): v1–v95
2. Figarella-Branger D et al (2021) The 2021 WHO classification of tumours of the central nervous system. Ann Pathol
3. Louis DN et al (2016) The 2016 World Health Organization classification of tumors of the central nervous system: a summary. Acta Neuropathol 131(6):803–820
4. Louis DN et al (2021) The 2021 WHO classification of tumors of the central nervous system: a summary. Neuro-Oncology 23(8):1231–1251
5. Plant-Fox AS, O'Halloran K, Goldman S (2021) Pediatric brain tumors: the era of molecular diagnostics, targeted and immune-based therapeutics, and a focus on long term neurologic sequelae. Curr Probl Cancer 45: 100777
6. Merker JD et al (2018) Circulating tumor DNA analysis in patients with cancer: American Society of Clinical Oncology and College of American Pathologists Joint Review. J Clin Oncol 36(16):1631–1641
7. Sweet-Cordero EA, Biegel JA (2019) The genomic landscape of pediatric cancers: implications for diagnosis and treatment. Science 363(6432):1170–1175
8. Panditharatna E et al (2018) Clinically relevant and minimally invasive tumor surveillance of pediatric diffuse midline gliomas using patient-derived liquid biopsy. Clin Cancer Res 24(23):5850–5859
9. Miller AM et al (2019) Tracking tumour evolution in glioma through liquid biopsies of cerebrospinal fluid. Nature 565(7741): 654–658
10. Miller AM et al (2022) Next-generation sequencing of cerebrospinal fluid for clinical molecular diagnostics in pediatric, adolescent and young adult (AYA) brain tumor patients. Neuro-Oncology 24:1763
11. De Mattos-Arruda L et al (2015) Cerebrospinal fluid-derived circulating tumour DNA better represents the genomic alterations of brain tumours than plasma. Nat Commun 6:8839
12. Wang Y et al (2015) Detection of tumor-derived DNA in cerebrospinal fluid of patients with primary tumors of the brain and spinal cord. Proc Natl Acad Sci USA 112(31): 9704–9709
13. Cantor E et al (2022) Serial H3K27M cell-free tumor DNA (cf-tDNA) tracking predicts ONC201 treatment response and progression in diffuse midline glioma. Neuro-Oncology 24(8):1366–1374
14. Liu APY et al (2021) Serial assessment of measurable residual disease in medulloblastoma liquid biopsies. Cancer Cell 39(11): 1519–1530 e4
15. Christodoulou E et al (2023) Combined low-pass whole genome and targeted sequencing in liquid biopsies for pediatric solid tumors. NPJ Precis Oncol 7(1):21
16. O'Halloran K et al (2023) Low-pass whole-genome and targeted sequencing of cell-free DNA from cerebrospinal fluid in pediatric patients with central nervous system tumors. Neurooncol Adv 5(1):vdad077
17. Adalsteinsson VA et al (2017) Scalable whole-exome sequencing of cell-free DNA reveals high concordance with metastatic tumors. Nat Commun 8(1):1324
18. Scheffler K et al (2023) Somatic small-variant calling methods in Illumina DRAGEN™ secondary analysis. bioRxiv:2023.03.23.534011
19. Robinson JT et al (2011) Integrative genomics viewer. Nat Biotechnol 29(1):24–26

Chapter 5

Modeling of Proton Radiotherapy Optimization for Pediatric Brain Tumors

Devin Miles, Lingshu Yin, and Heng Li

Abstract

This chapter describes the planning aspects of proton beam therapy for pediatric brain malignancies, focusing on optimizing proton radiotherapy through improved modeling and optimization of relative biological effectiveness (RBE) and linear energy transfer (LET). Proton therapy offers potential advantages over conventional photon therapy owing to its depth dose characteristics, including the Bragg peak, the limited exit dose, and the minimized integral dose, thus making it particularly appealing for pediatric patients. Various proton RBE models, including the Local Effect Model (LEM), Microdosimetric Kinetic Model (MKM), and Repair-Misrepair Fixation Model (RMF), and approaches to RBE/LET optimization in treatment planning are discussed. Understanding and incorporating LET and RBE considerations into proton therapy planning may lead to improved treatment outcomes and reduced long-term toxicities for pediatric CNS tumors.

Key words Radiotherapy, Protons, LET optimization, Pediatric brain tumor

1 Introduction

Radiation therapy is often the primary treatment modality for pediatric central nervous system (CNS) tumors, Hodgkin's lymphoma, neuroblastoma, medulloblastoma, and bone neoplasms [1, 2]. It plays a crucial role in treating these cancers by directly targeting the tumor site to achieve local control and disease eradication. Technological advancements, such as image guidance and intensity modulation, have increased the precision of tumor targeting, reducing error margins and minimizing radiation exposure to healthy tissues [3, 4]. These advancements have led to improved local control and survival rates for pediatric brain and CNS cancers, with 5-year overall survival rates exceeding 70% [5].

Pediatric patients, given their heightened radiation susceptibility and longer life expectancy, are particularly vulnerable to the short-term and long-term secondary effects of radiation therapy

Michael Goldstein (ed.), *Translational Research Methods for Pediatric Brain Tumors*, Neuromethods, vol. 226,
https://doi.org/10.1007/978-1-0716-4754-7_5,

[6]. Radiation exposure to the brain can result in numerous cognitive and functional deficits, including cognitive impairment, hearing loss, endocrine dysfunction, and a heightened risk of secondary cancers [7]. Therefore, optimal radiation therapy planning for pediatric patients should thoroughly consider survival outcomes, radiation therapy-associated toxicities, and strategies to mitigate both acute and late effects, while prioritizing local control [8].

In radiation oncology clinics, high-energy X-ray photon beams generated by linear accelerators are commonly used for external beam radiation therapy [9]. These therapeutic megavoltage photon beams are characterized by an energy-dependent buildup and a shallow depth-dose falloff. Linear accelerators utilize onboard imaging and variable beam-shaping tools, such as multi-leaf collimators, to design highly conformal image-guided radiation deliveries to a target. Techniques such as intensity-modulated radiation therapy (IMRT) or volumetric modulated arc therapy (VMAT) are employed. However, due to the presence of an "exit dose"—the photon dose downstream of a target—advanced photon beam therapy can result in a substantial volume of healthy, uninvolved tissue receiving a low-to-moderate dose of radiation [10].

Proton beam therapy, though less common than photon beam therapy and available at only 45 centers in the US at the time of writing [11], is a growing modality of radiation therapy that has been highly successful in treating pediatric cancers. Proton therapy leverages the Bragg peak phenomenon, where protons undergo a higher density of physical interactions near their end of the range, resulting in a steep increase ('peak') in dose deposition before stopping [12, 13]. The depth of the Bragg peak is energy-dependent, allowing proton beams to deliver high doses of radiation at a precisely calculated depth with no residual exit dose. The width of the Bragg peak can be modulated by stacking proton beams of different energies, creating a spread-out Bragg peak (SOBP). Energy modulation can be performed using either a degrader in a cyclotron-based system or by selectively accelerating protons to a desired energy in a synchrotron-based system. Additionally, the shaping of a proton beam perpendicular to the beam axis can be done using double-scatter systems with ancillary collimation or by using pencil beam scanning [14]. The state-of-the-art in proton beam therapy is pencil beam scanning, where a narrow proton beam is magnetically steered to "paint" a dose perpendicular to the beam axis. This precise control of proton beams can reduce the volume of healthy tissues receiving radiation, thereby lowering the risk of long-term toxicities and secondary primary cancer development compared to X-ray radiation therapy [15, 16].

While proton beams can offer preferable dose distributions for pediatric patients with CNS cancers, proton doses are not always directly comparable to photon doses due to differences in radiation quality and patterns of DNA damage and repair [17]. These differences complicate cross-modality comparisons and confound the application of X-ray radiation therapy research findings to proton therapy. The concept of relative biological effectiveness (RBE) is used to quantify the varying effectiveness of different types of radiation in producing biological damage. Proton therapy treatments are typically based on a uniform proton RBE relative to high-energy photons of around 1.1, which represents the average effectiveness at inducing DNA damage at the center of an SOBP [18]. However, it is now known that proton and other ion beams exhibit an increasing RBE with increasing depth toward the distal edge of the Bragg peak [19]. RBE could also vary with cell types, end point, and dose. The varying RBE raises questions and potential opportunities to adjust dose prescriptions and organ-at-risk constraints during treatment planning [20]. For many pediatric applications of proton therapy, the high-RBE distal edge may be placed downstream of the target, resulting in the most highly damaging region of the proton beam affecting healthy at-risk structures [21].

Recent advances in mainstream treatment planning tools have enabled the incorporation of particle linear energy transfer (LET) and variable RBE models in optimization, but guidance for their optimal implementation in the clinic is still being developed [19]. The following sections will describe the concepts of LET and RBE in greater detail and introduce strategies for considering these metrics in proton treatment planning for pediatric brain malignancies.

2 Materials

1. Concepts Describing Linear Energy Transfer (LET) and Relative Biological Effectiveness (RBE).

2.1 Linear Energy Transfer (LET)

Linear energy transfer (LET) describes the amount of energy deposited by ionizing radiation per unit distance traveled in a medium, typically expressed in units of keV per micrometer (keV/μm) [18]. High-LET radiation deposits more energy within a shorter distance along its path, leading to dense patterns of ionization interactions and more complex biological damage compared to low-LET radiation [22, 23]. This increased biological effectiveness is attributed to high-LET radiation's ability to induce clustered DNA damage that is challenging for cells to repair, making it more effective in killing tumor cells and overcoming radioresistance [24].

While direct physical measurement of LET is not currently achievable, it can be experimentally measured using various techniques. Historically, gas-based tissue-equivalent proportional counters were used to measure the distribution of energy depositions in a small effective volume, comparable to the diameter of a single cell [25, 26]. This approach was fundamental to microdosimetry, which aimed to characterize the stochastic nature of radiation interactions within cells [27]. Microdosimetric analogs to deposited dose (specific energy) and LET (lineal energy) have been used in various models of relative biological effectiveness (RBE), which will be discussed in detail in a subsequent section.

Modern "microdosimeters" use solid-state detectors to quantify the energy deposited by radiation along the tracks of charged particles at microscopic and nanoscopic scales [28, 29]. These detectors provide information on the linear energy spectra at different depths and positions in a phantom, allowing for verification of radiation quality in hadron therapy. More commonly available clinical detectors, such as radiochromic film [30] or optically stimulated luminescent dosimeters [31], are used for indirect measurement of LET. Monte Carlo simulations, which track the energy deposition and interaction patterns of individual ionizing radiation particles, are often used to model LET in a medium [32]. Various Monte Carlo packages, such as GEANT4 [33], MCNP [34], FLUKA [35], and TOPAS [36], have been applied to model LET in the context of therapeutic proton beams. Recent developments in clinical treatment planning software have incorporated Monte Carlo dose engines, allowing the simulation of LET directly in the planning interface [37, 38]. However, no currently available treatment planning systems (TPS) have built-in functionality to optimize based on LET, although this concept has been explored by many investigators [39]. When modeling LET, individual particles deposit variable energy per interaction, with varying separation between ionization interactions, and averages in LET are often expressed on a macroscopic voxel-by-voxel basis.

Dose-averaged LET (LETd) and track-averaged LET (LETt) are two approaches to modeling and reporting LET on a macroscopic level [32, 40, 41]. LETd is calculated by weighting the LET of each particle by its contribution to the local dose, often interpreted as a metric of additional dose due to LET effects. In contrast, LETt is calculated by weighting the overall LET by the local fluence of particles at a given LET, providing insight into the average local energy deposition patterns and variations within cellular or subcellular scales.

LETd is more commonly applied for modeling the biological effectiveness of radiation therapy due to a frequently observed linear correlation between LETd and RBE [18]. However, this may not accurately represent the intricate biological responses to radiation, as RBE values can vary based on factors beyond LET alone [42]. Additional parameters, such as the weighting in LET averaging (track-averaging vs. dose-averaging), voxel resolution, energy transport cutoffs, and the selection of considered secondary

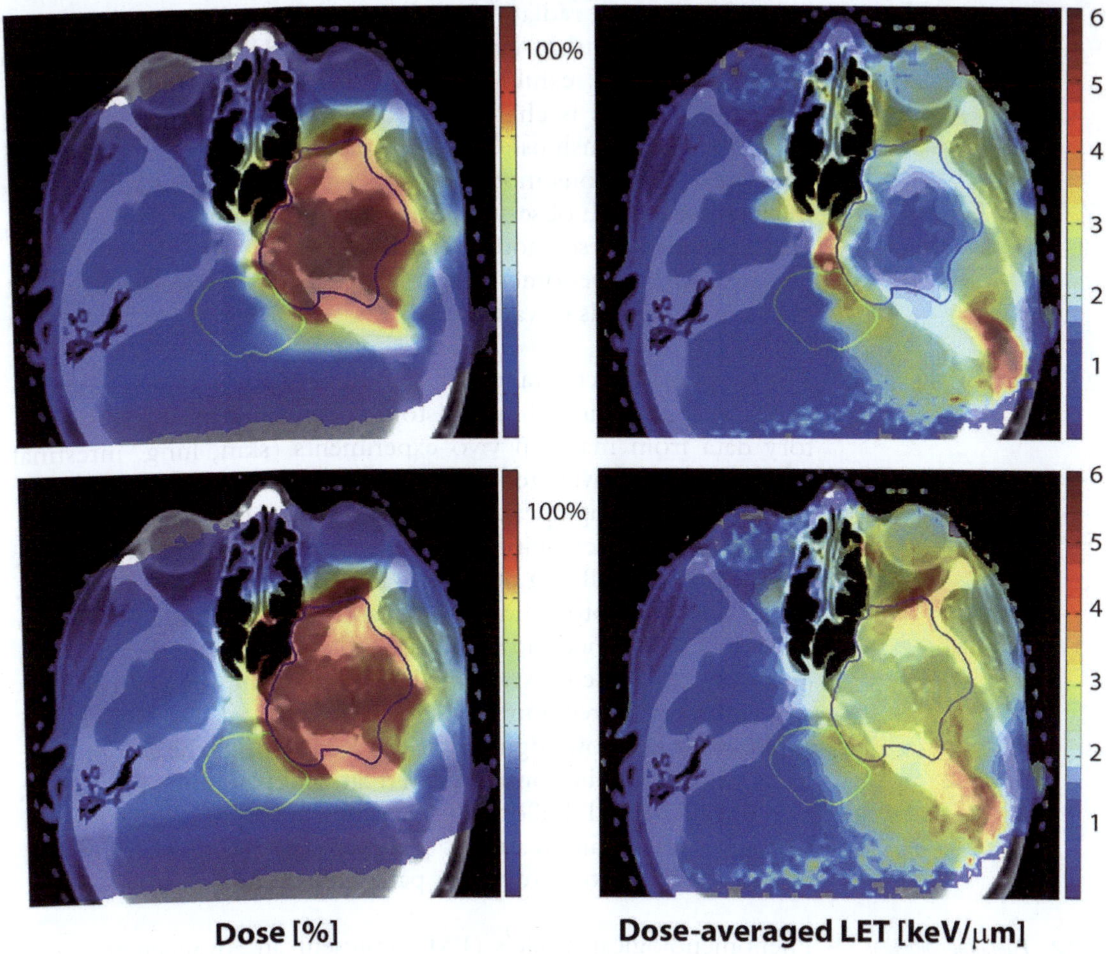

Fig. 1 Dose (in %) and dose-averaged LET (in keV/mm) for two IMPT treatment plans. (Figure from Paganetti et al., NRG Oncology White Paper on the Relative Biological Effectiveness in Proton Therapy. 2024; Int J Radiat Oncol Biol Phys, in press. Figure reproduced with permission, license number 5930730239797)

particles, can introduce uncertainties in LET averaging methods and reporting, impacting treatment planning and optimization strategies [43].

Figure 1 shows the dose and dose-averaged LET in keV/mm for two IMPT treatment plans. Even though the dose distributions are similar, the LETd distributions are substantially different between plans, which could lead to different radiation responses. Thus, strategic incorporation of LETd into proton RBE modeling is critical in evaluating and optimization of proton treatment plans.

2.2 Models of Variable Relative Biological Effectiveness (RBE)

Relative biological effectiveness (RBE) quantifies the effectiveness of a given type of radiation in producing a biological outcome compared to a reference radiation, typically X-rays or gamma rays. RBE values are influenced by various factors, including LET, dose, specific endpoints, radiation track structure, and target tissue type [44]. The relationship between LET and RBE is intricate, with different particles exhibiting varying RBE-LET curves. Direct modeling of RBE is challenging due to uncertainties associated with the relationships between dose, LET, and biological responses. Numerous models have been developed to account for the increase in RBE observed at the distal edge of the proton Bragg peak. Broadly, these models can be stratified into three groups: uniform RBE, phenomenological models of variable RBE, and mechanistic models of variable RBE.

2.2.1 Uniform RBE

The adoption of a constant RBE value in proton therapy is based on practical considerations and historical conventions. Early laboratory data from many in vivo experiments (skin, lung, intestinal crypt) showed an average RBE of 1.10 (range 0.7–1.5) at the center of a spread-out Bragg peak [45]. In vitro data, such as for V79 cells, allowed greater flexibility to sample the more highly damaging distal edge of the Bragg peak, with an RBE of 1.4 [46]. However, at the time of proton therapy's expansion in the early 2000s, no clear clinical evidence had demonstrated a comparable finding in an in vivo system. The most recent ICRU guidelines on proton therapy continue to recommend a generic RBE of 1.1 for clinical applications, acknowledging that elevated RBE at the Bragg peak distal edge should be considered in treatment planning [47]. Using a uniform RBE of 1.1 allows for the simple incorporation of proton biological effects in treatment planning and the straightforward conversion of photon treatment paradigms to proton therapy.

2.2.2 Phenomenological Models

Phenomenological models (PM) represent an advancement over the uniform RBE by considering proton biological effects under varying conditions. Developed from empirical clinical or laboratory data, PMs estimate trends or patterns of biological damage from proton radiotherapy for specific endpoints or cell types [48]. These models often modify the linear-quadratic (LQ) model of radiation dose response, applying scaling functions to photon alpha and beta LQ parameters based on factors like proton LET [49].

Numerous PMs exist to estimate proton RBE, with approximately 13 included in a recent review article by McMahon (2021) [50]. Commonly implemented models include the Wedenberg [51], Carabe-Fernandez [52], and McNamara [53] models. Despite a positive correlation among models, there is significant variability due to differences in how they account for photon alpha/beta ratios in calculating proton alpha or proton RBE-max. Different weightings in the scaling of proton alpha/beta can result in substantial differences between models, potentially confounding RBE prediction in clinically relevant scenarios [50]. There is no clear consensus on the "optimal" phenomenological model, as they seek to fit data to the proton biological response under specific experimental conditions. Future systematic evaluations of proton RBE in various cell types and well-controlled experimental conditions may help unify individual PMs and enable their use in proton radiotherapy planning with greater confidence.

2.2.3 Mechanistic RBE Models

Mechanistic RBE models further advance the understanding of RBE by explicitly considering molecular and cellular radiation responses [20]. These models encompass various aspects of the biological response to radiation damage, including the spatial distribution and frequency of DNA double-strand breaks (DSB) induction, DNA repair mechanisms, intercellular interactions, tissue oxygenation, and the induction and scavenging of free radicals. By integrating these factors, mechanistic models explore how different radiation qualities can impact biological outcomes. Three mainstream mechanistic RBE models exist for proton and other ion therapy applications: the Local Effect Model (LEM), the Microdosimetric Kinetic Model (MKM), and the Repair-Misrepair Fixation Model (RMF).

Local Effect Model (LEM)

The Local Effect Model (LEM) is a mathematical model to calculate the biological outcomes of ionizing radiation by focusing on the spatial distribution of energy deposition and the formation of DNA double-strand breaks (DSBs) along the paths of charged particles [54, 55]. The model assumes that the biological effect at any given location within the cell nucleus depends on the local dose and the local energy density of ionizing events [56]. This approach allows LEM to provide a nuanced prediction of cell survival.

LEM transitions from a linear-quadratic (LQ) to a linear-quadratic-linear (LQL) dose-response model for higher doses, reflecting the increased effectiveness of high-linear energy transfer (LET) radiation in inducing clustered DNA damage [57]. Such clustered damage is more difficult for cells to repair, making high-LET radiation more biologically effective at killing tumor cells and overcoming radioresistance [22]. The model incorporates detailed information about the track structure of various types of radiation, including protons and heavier ions like carbon, to accurately describe the RBE-LET relationship for different ion species. This capability is particularly important in proton and ion therapy due to LET enhancement at the distal edge of the Bragg peak. Although LEM offers a robust method for optimizing dose distributions in particle therapy, its application is complex, requiring extensive input data on radiation track structure and biological responses.

Microdosimetric Kinetic Model (MKM)

The Microdosimetric Kinetic Model focuses on the stochastic nature of energy deposition within cellular volumes, considering individual ionization interactions and their capability to induce critical biological damage like DSBs [58–60]. One key distinction of the MKM is its consideration of the microdosimetric spectra, which detail the distribution of energy deposited in small volumes, such as the dimensions of the cell nucleus. These spectra reflect the variability and intensity of ionization events that can contribute to DNA damage. The model incorporates the concept of LET (via the metric of lineal energy), where higher LET values correspond to denser ionization tracks that cause more clustered DNA damage, making radiation more effective at killing tumor cells but also potentially more damaging to healthy tissues.

The MKM also integrates knowledge of DNA repair kinetics, highlighting how cells respond to and attempt to repair radiation-induced damage. This mechanistic approach allows the MKM to predict cell survival and other biological endpoints based on detailed simulations of energy deposition patterns and subsequent biological responses.

Repair-Misrepair Fixation (RMF) Model

The Repair-Misrepair Fixation (RMF) model distinguishes itself by incorporating detailed mechanisms of DNA repair and chromatin interactions to predict cellular responses to radiation-induced damage [61, 62]. The RMF model proposes a dual-component response mechanism, where initial DSBs are subject to repair processes within the cell nucleus. The RMF model delineates these repair processes into a first-order repair term, representing the typical exponential decay of DSBs as repair progresses, and a second-order (quadratic) term that accounts for a smaller subset of DSBs undergoing complex interactions leading to misrepair or formation of chromosomal exchanges.

Unlike other models, the RMF model considers the entire nuclear volume as a plausible site for the interactions responsible for misrepair, emphasizing the significance of chromatin structure and repair kinetics in determining RBE values.

It is worth noting that because of the fundamental difference in assumptions for these mechanistic models, they require vastly different input parameters and could produce substantially different results, as demonstrated in Fig. 2 [20]. As such, the authors recommended further comparisons and validations of model predictions with experimental data among RBE models of particle.

3 Methods

3.1 RBE/LET Optimization

In addition to more accurate RBE/LET optimization for radiotherapy of pediatric brain tumors, it has become an important area of research, aiming to improve treatment outcomes while minimizing side effects in these vulnerable patients.

One key method involves the use of variable RBE models instead of the conventional fixed RBE of 1.1 used in clinical proton therapy. These variable RBE models, such as the McNamara model

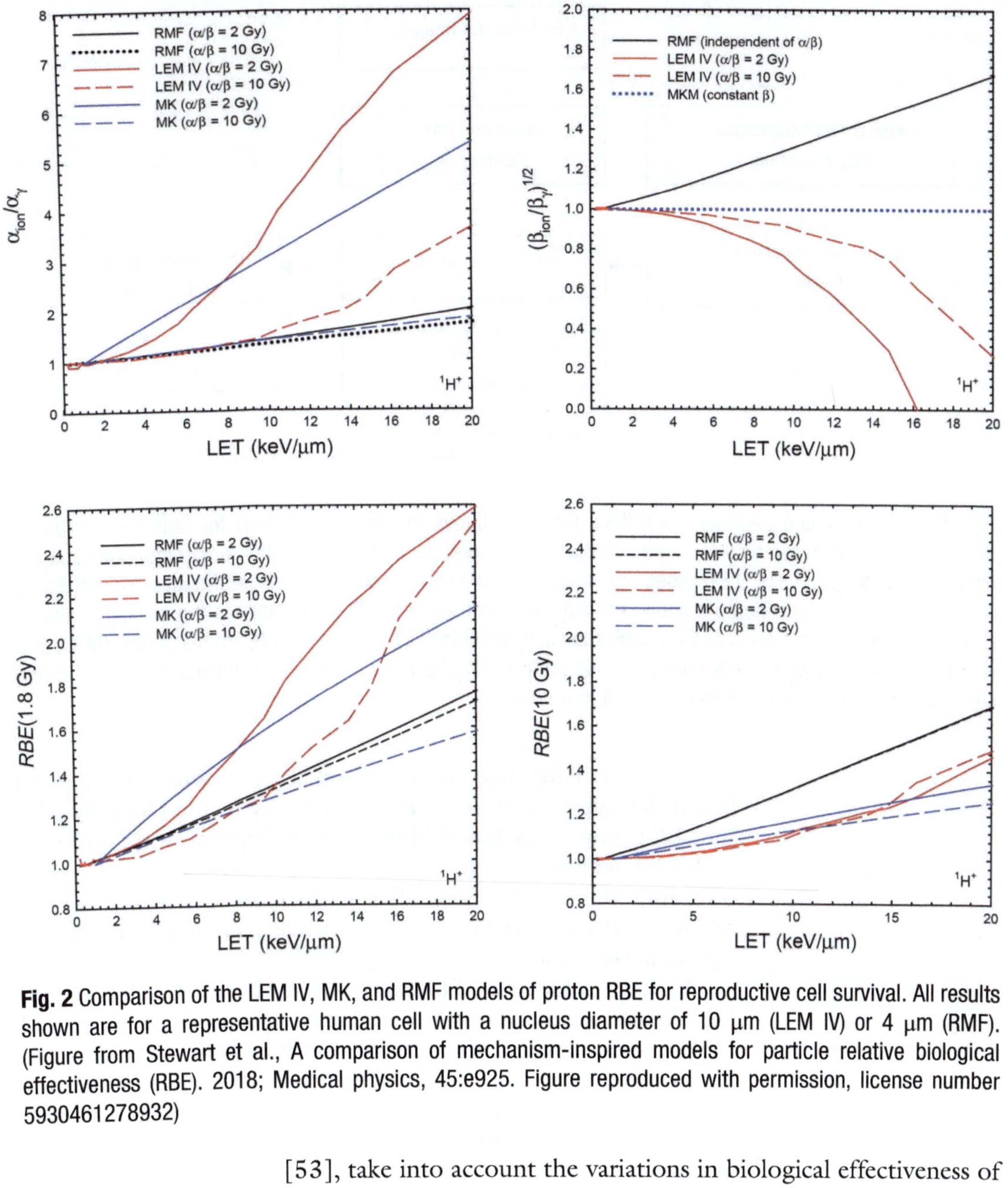

Fig. 2 Comparison of the LEM IV, MK, and RMF models of proton RBE for reproductive cell survival. All results shown are for a representative human cell with a nucleus diameter of 10 μm (LEM IV) or 4 μm (RMF). (Figure from Stewart et al., A comparison of mechanism-inspired models for particle relative biological effectiveness (RBE). 2018; Medical physics, 45:e925. Figure reproduced with permission, license number 5930461278932)

[53], take into account the variations in biological effectiveness of protons at different depths and energies. By incorporating these models into treatment planning, clinicians can better predict and account for the increased biological effect of protons, especially at the end of their range where LET is highest.

Figure 3 shows the typical workflow for radiation treatment. RBE/LET optimization can be integrated into the typical workflow in the optimization parameter selection and plan optimization process. LET-based optimization has been proposed as an

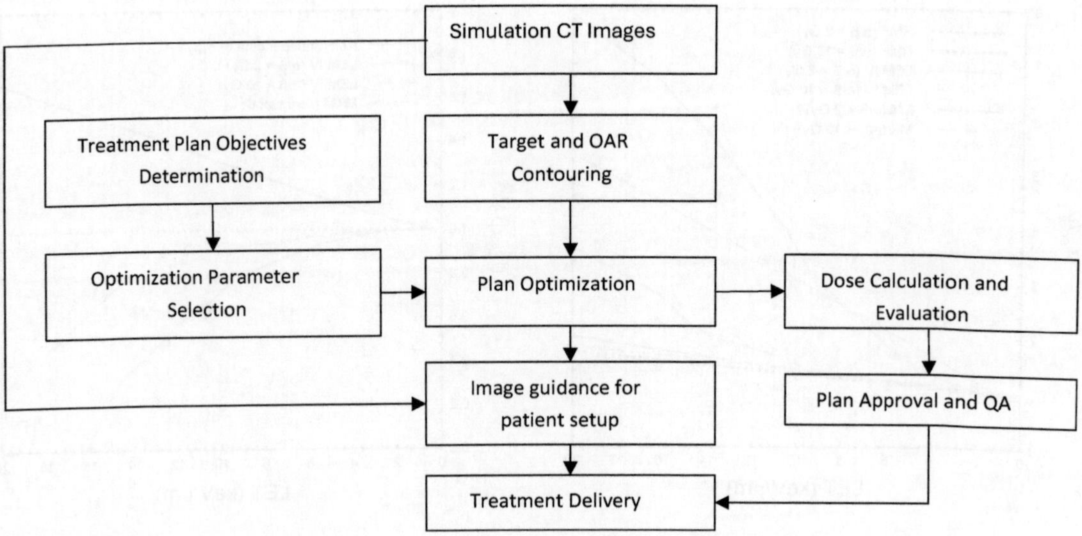

Fig. 3 Typical radiation treatment workflow. Figure 3 shows the typical workflow for radiation treatment. RBE/LET optimization can be integrated into the typical workflow in the optimization parameter selection and plan optimization process. LET-based optimization has been proposed as an alternative to RBE-based optimization. LET is considered a good first-order approximation of RBE, especially in the clinically relevant LET range. The general intent of LET optimization is to shift high LET regions outside organs at risk (OARs) and, where possible, inside the target volume. This method aims to reduce the risk of adverse effects in critical structures while maintaining or improving tumor control

alternative to RBE-based optimization. LET is considered a good first-order approximation of RBE, especially in the clinically relevant LET range. The general intent of LET optimization is to shift high LET regions outside organs at risk (OARs) and, where possible, inside the target volume. This method aims to reduce the risk of adverse effects in critical structures while maintaining or improving tumor control.

Various objective functions have been developed for LET-based optimization. Some studies use quadratic dose and LET objectives, while others employ linear or nonlinear objectives. For example, Cao et al. [63] and Chen et al. [64] applied a one-step linear objective function to five brain tumor cases, demonstrating the feasibility of this approach in pediatric brain tumor treatment planning.

Robustness analysis is another crucial aspect of RBE/LET optimization for pediatric brain tumors [65, 66]. This involves evaluating the sensitivity of treatment plans to physical uncertainties such as patient setup errors and range uncertainties. Robust optimization techniques can be employed to create plans that maintain target coverage and OAR sparing across various uncertainty scenarios, ensuring more reliable treatment delivery.

Clinical evidence supporting the importance of considering variable RBE and LET in treatment planning is emerging. A study by Peeler et al. [67] found that areas of normal tissue damage indicated by post-treatment image changes in pediatric ependymoma patients were associated with increased biological dose effectiveness. They observed that the dose at which image changes occurred was lower when combined with elevated LET values, suggesting an increase in biological dose effectiveness with increased LET.

It's important to note that while these optimization methods show promise, there is still a need for further clinical evidence to fully understand their impact. Researchers emphasize the importance of careful clinical and imaging follow-up, as well as continued investigation into end-of-range RBE mitigation strategies for pediatric brain tumor patients treated with proton therapy.

In conclusion, RBE/LET optimization methods for radiotherapy of pediatric brain tumors are evolving rapidly. These techniques offer the potential to improve treatment outcomes by accounting for the variable biological effectiveness of protons and optimizing LET distributions. However, ongoing research is needed to refine these methods and establish their clinical benefits in this sensitive patient population.

4 Notes

Understanding RBE is critical in proton therapy as it describes how effectively biological damage is induced compared to conventional X-ray radiation therapy. RBE values vary based on factors like LET, dose, specific endpoints, radiation track structure, and tissue type. The intricate relationship between LET and RBE necessitates accurate modeling to optimize proton therapy outcomes [68].

The numerous models of proton RBE provide essential frameworks for this purpose. Uniform RBE models simplify treatment planning but may overlook nuances in biological response, particularly in LET-enhanced regions of a proton beam. Phenomenological models can adapt to different conditions, refining predictions based on empirical data. Mechanistic models delve into molecular and cellular responses, offering detailed insights into radiation-induced damage mechanisms. These models allow estimation and prediction of cell survival and biological responses, as well as support the optimization of proton therapy strategies. By accounting for complex factors influencing RBE, these advancements in modeling can contribute to improving therapeutic efficacy while minimizing risks to healthy tissues.

In summary, understanding the impact of LET and RBE in proton therapy and incorporating them into treatment planning are critical for pediatric brain tumors. The transition from a uniform

RBE to variable RBE models and LET-based optimization strategies represents a significant advancement in the field, and may lead to improved treatment outcomes, including reduced long-term toxicities. However, further research is needed to facilitate the optimal implementation of these concepts in clinical practice.

Conflict of Interest

The authors have no conflicts of interest to declare that are relevant to the content of this chapter.

References

1. Merchant TE (2009) Proton beam therapy in pediatric oncology. Cancer J 15(4):298. https://doi.org/10.1097/PPO.0b013e3181b6d4b7
2. Merchant TE (2013) Clinical controversies: proton therapy for pediatric tumors. Semin Radiat Oncol 23(2):97–108. https://doi.org/10.1016/j.semradonc.2012.11.008
3. Lane SA, Slater JM, Yang GY (2023) Image-guided proton therapy: a comprehensive review. Cancers 15(9):2555. https://doi.org/10.3390/cancers15092555
4. Moreno AC, Frank SJ, Garden AS et al (2019) Intensity modulated proton therapy (IMPT)—the future of IMRT for head and neck cancer. Oral Oncol 88:66–74. https://doi.org/10.1016/j.oraloncology.2018.11.015
5. Bhatia S, Sklar C (2002) Second cancers in survivors of childhood cancer. Nat Rev Cancer 2(2):124–132. https://doi.org/10.1038/nrc722
6. Hall EJ (2006) Intensity-modulated radiation therapy, protons, and the risk of second cancers. Int J Radiation Oncol Biol Phy 65(1):1–7. https://doi.org/10.1016/j.ijrobp.2006.01.027
7. Lundkvist J, Ekman M, Ericsson SR, Jönsson B, Glimelius B (2005) Cost-effectiveness of proton radiation in the treatment of childhood medulloblastoma. Cancer 103(4):793–801. https://doi.org/10.1002/cncr.20844
8. DeNunzio NJ, Yock TI (2020) Modern radiotherapy for pediatric brain tumors. Cancers. 12(6):1533. https://doi.org/10.3390/cancers12061533
9. Khan FM (2003) The physics of radiation therapy, 3rd edn. Lippincott Williams & Wilkins
10. Levin WP, Kooy H, Loeffler JS, DeLaney TF (2005) Proton beam therapy. Br J Cancer 93(8):849–854. https://doi.org/10.1038/sj.bjc.6602754
11. Find A Proton Center. Accessed 19 Nov 2024. https://proton-therapy.org/findacenter/
12. Mohan R (2022) A review of proton therapy—current status and future directions. Precis Radiat Oncol 6(2):164–176. https://doi.org/10.1002/pro6.1149
13. Newhauser WD, Zhang R (2015) The physics of proton therapy. Phys Med Biol 60(8):R155–R209. https://doi.org/10.1088/0031-9155/60/8/R155
14. Lomax A (1999) Intensity modulation methods for proton radiotherapy. Phys Med Biol 44(1):185. https://doi.org/10.1088/0031-9155/44/1/014
15. Chuong M, Badiyan SN, Yam M et al (2018) Pencil beam scanning versus passively scattered proton therapy for unresectable pancreatic cancer. J Gastrointest Oncol 9(4):687–693. https://doi.org/10.21037/jgo.2018.03.14
16. Arjomandy B, Sahoo N, Cox J, Lee A, Gillin M (2009) Comparison of surface doses from spot scanning and passively scattered proton therapy beams. Phys Med Biol 54(14):N295. https://doi.org/10.1088/0031-9155/54/14/N02
17. Valentin J (2003) Relative biological effectiveness (RBE), quality factor (Q), and radiation weighting factor (wR): ICRP publication 92: approved by the Commission in January 2003. Ann ICRP 33(4):1–121. https://doi.org/10.1016/S0146-6453(03)00024-1
18. Paganetti H (2014) Relative biological effectiveness (RBE) values for proton beam therapy. Variations as a function of biological endpoint, dose, and linear energy transfer. Phys Med Biol 59(22):R419–R472. https://doi.org/10.1088/0031-9155/59/22/R419
19. Paganetti H, Simone CB, Bosch WR et al (2024) NRG oncology white paper on the

relative biological effectiveness in proton therapy. Int J Radiation Oncol Biol Phy. Published online July 25, 2024. https://doi.org/10.1016/j.ijrobp.2024.07.2152

20. Stewart RD, Carlson DJ, Butkus MP, Hawkins R, Friedrich T, Scholz M (2018) A comparison of mechanism-inspired models for particle relative biological effectiveness (RBE). Med Phys 45(11):e925. https://doi.org/10.1002/mp.13207
21. Ladra MM, MacDonald SM, Terezakis SA (2018) Proton therapy for central nervous system tumors in children. Pediatr Blood Cancer 65(7):e27046. https://doi.org/10.1002/pbc.27046
22. Nikjoo H, Emfietzoglou D, Liamsuwan T, Taleei R, Liljequist D, Uehara S (2016) Radiation track, DNA damage and response—a review. Rep Prog Phys 79(11):116601. https://doi.org/10.1088/0034-4885/79/11/116601
23. Baskar R, Dai J, Wenlong N, Yeo R, Yeoh KW (2014) Biological response of cancer cells to radiation treatment. Front Mol Biosci 1:1. https://doi.org/10.3389/fmolb.2014.00024
24. Wank M, Schilling D, Reindl J et al (2018) Evaluation of radiation-related invasion in primary patient-derived glioma cells and validation with established cell lines: impact of different radiation qualities with differing LET. J Neuro-Oncol 139(3):583–590. https://doi.org/10.1007/s11060-018-2923-4
25. Rossi HH, Zaider M (1996) Microdosimetry and its applications. Springer, Berlin/Heidelberg. https://doi.org/10.1007/978-3-642-85184-1
26. Luxton G, Fessenden P, Hoffmann W (1979) Microdosimetric measurements of Pretherapeutic heavy ion beams. Radiat Res 79(2):256–272. https://doi.org/10.2307/3575096
27. Kellerer AM (1985) Fundamentals of microdosimetry. In: The dosimetry of ionizing radiation. Elsevier, pp 77–162. https://doi.org/10.1016/B978-0-12-400401-6.50007-3
28. Anderson SE, Furutani KM, Tran LT et al (2017) Microdosimetric measurements of a clinical proton beam with micrometer-sized solid-state detector. Med Phys 44(11):6029–6037. https://doi.org/10.1002/mp.12583
29. Tran LT, Bolst D, Guatelli S et al (2018) The relative biological effectiveness for carbon, nitrogen, and oxygen ion beams using passive and scanning techniques evaluated with fully 3D silicon microdosimeters. Med Phys 45(5):2299–2308. https://doi.org/10.1002/mp.12874
30. Kawashima M, Matsumura A, Souda H, Tashiro M (2020) Simultaneous determination of the dose and linear energy transfer (LET) of carbon-ion beams using radiochromic films. Phys Med Biol 65(12):125002. https://doi.org/10.1088/1361-6560/ab8bf3
31. Sawakuchi GO, Sahoo N, Gasparian PBR et al (2010) Determination of average LET of therapeutic proton beams using Al_2O_3: C optically stimulated luminescence (OSL) detectors. Phys Med Biol 55(17):4963–4976. https://doi.org/10.1088/0031-9155/55/17/006
32. Bertolet A, Baratto-Roldán A, Barbieri S, Baiocco G, Carabe A, Cortés-Giraldo M, a. (2019) Dose-averaged LET calculation for proton track segments using microdosimetric Monte Carlo simulations. Med Phys 46(9):4184–4192. https://doi.org/10.1002/mp.13643
33. Grassberger C, Trofimov A, Lomax A, Paganetti H (2011) Variations in linear energy transfer within clinical proton therapy fields and the potential for biological treatment planning. Int J Radiation Oncol Biol Phy 80(5):1559–1566. https://doi.org/10.1016/j.ijrobp.2010.10.027
34. Lillhök J, Billnert-Maróti R, Anastasiadis A (2022) MCNP 6.2 simulations of energy deposition in low-density volumes corresponding to unit-density volumes on the nanometre level. Radiat Meas 157:106831. https://doi.org/10.1016/j.radmeas.2022.106831
35. Fjæra LF, Li Z, Ytre-Hauge KS et al (2017) Linear energy transfer distributions in the brainstem depending on tumour location in intensity-modulated proton therapy of paediatric cancer. Acta Oncol 56(6):763–768. https://doi.org/10.1080/0284186X.2017.1314007
36. Granville DA, Sawakuchi GO (2015) Comparison of linear energy transfer scoring techniques in Monte Carlo simulations of proton beams. Phys Med Biol 60(14):N283. https://doi.org/10.1088/0031-9155/60/14/N283
37. Vaniqui A, Vaassen F, Di Perri D et al (2023) Linear energy transfer and relative biological effectiveness investigation of various structures for a cohort of proton patients with brain tumors. Adv Radiat Oncol 8(2):101128. https://doi.org/10.1016/j.adro.2022.101128
38. Janson M, Glimelius L, Fredriksson A, Traneus E, Engwall E (2024) Treatment planning of scanned proton beams in

RayStation. Med Dosim 49(1):2–12. https://doi.org/10.1016/j.meddos.2023.10.009

39. Deng W, Yang Y, Liu C et al (2021) A critical review of LET-based intensity-modulated proton therapy plan evaluation and optimization for head and neck cancer management. Int J Part Ther 8(1):36–49. https://doi.org/10.14338/IJPT-20-00049.1

40. Smith EAK, Winterhalter C, Underwood TSA et al (2022) A Monte Carlo study of different LET definitions and calculation parameters for proton beam therapy. Biomed Phys Eng Express 8(1):015024. https://doi.org/10.1088/2057-1976/ac3f50

41. Guan F, Peeler C, Bronk L et al (2015) Analysis of the track- and dose-averaged LET and LET spectra in proton therapy using the GEANT 4 Monte Carlo code. Med Phys 42(11):6234–6247. https://doi.org/10.1118/1.4932217

42. Grün R, Friedrich T, Traneus E, Scholz M (2019) Is the dose-averaged LET a reliable predictor for the relative biological effectiveness? Med Phys 46(2):1064–1074. https://doi.org/10.1002/mp.13347

43. Hahn C, Ödén J, Dasu A et al (2022) Towards harmonizing clinical linear energy transfer (LET) reporting in proton radiotherapy: a European multi-centric study. Acta Oncol 61(2):206–214. https://doi.org/10.1080/0284186X.2021.1992007

44. Gulliford SL, Prise KM (2019) Relative biological effect/linear energy transfer in proton beam therapy: a primer. Clin Oncol 31(12):809–812. https://doi.org/10.1016/j.clon.2019.06.009

45. Paganetti H, Niemierko A, Ancukiewicz M et al (2002) Relative biological effectiveness (RBE) values for proton beam therapy. Int J Radiation Oncol Biol Phy 53(2):407–421. https://doi.org/10.1016/S0360-3016(02)02754-2

46. Robertson JB, Williams JR, Schmidt RA, Little JB, Flynn DF, Suit HD (1975) Radiobiological studies of a high-energy modulated proton beam utilizing cultured mammalian cells. Cancer 35(6):1664–1677. https://doi.org/10.1002/1097-0142(197506)35:6<1664::AID-CNCR2820350628>3.0.CO;2-#

47. DeLuca PM (2007) The INTERNATIONAL COMMISSION on radiation units and measurements. J ICRU 7(2):NP–NP. https://doi.org/10.1093/jicru/ndn001

48. Rørvik E, Thörnqvist S, Stokkevåg CH, Dahle TJ, Fjæra LF, Ytre-Hauge KS (2017) A phenomenological biological dose model for proton therapy based on linear energy transfer spectra. Med Phys 44(6):2586–2594. https://doi.org/10.1002/mp.12216

49. Giovannini G, Böhlen T, Cabal G et al (2016) Variable RBE in proton therapy: comparison of different model predictions and their influence on clinical-like scenarios. Radiat Oncol 11(1):68. https://doi.org/10.1186/s13014-016-0642-6

50. McMahon SJ (2021) Proton RBE models: commonalities and differences. Phys Med Biol 66(4):04NT02. https://doi.org/10.1088/1361-6560/abda98

51. Wedenberg M, Lind BK, Hårdemark B (2013) A model for the relative biological effectiveness of protons: the tissue specific parameter α / β of photons is a predictor for the sensitivity to LET changes. Acta Oncol 52(3):580–588. https://doi.org/10.3109/0284186X.2012.705892

52. Carabe A, Moteabbed M, Depauw N, Schuemann J, Paganetti H (2012) Range uncertainty in proton therapy due to variable biological effectiveness. Phys Med Biol 57(5):1159–1172. https://doi.org/10.1088/0031-9155/57/5/1159

53. McNamara AL, Schuemann J, Paganetti H (2015) A phenomenological relative biological effectiveness (RBE) model for proton therapy based on all published *in vitro* cell survival data. Phys Med Biol 60(21):8399–8416. https://doi.org/10.1088/0031-9155/60/21/8399

54. Elsässer T, Scholz M (2007) Cluster effects within the local effect model. Radiat Res 167(3):319–329. https://doi.org/10.1667/RR0467.1

55. Friedrich T, Durante M, Scholz M. The local effect model—principles and applications. 14

56. Scholz M, Kellerer AM, Kraft-Weyrather W, Kraft G (1997) Computation of cell survival in heavy ion beams for therapy. Radiat Environ Biophys 36(1):59–66. https://doi.org/10.1007/s004110050055

57. Krämer M, Scholz M (2006) Rapid calculation of biological effects in ion radiotherapy. Phys Med Biol 51(8):1959–1970. https://doi.org/10.1088/0031-9155/51/8/001

58. Hawkins RB (1996) A microdosimetric-kinetic model of cell death from exposure to ionizing radiation of any LET, with experimental and clinical applications. Int J Radiat Biol 69(6):739–755. https://doi.org/10.1080/095530096145481

59. Hawkins RB, Inaniwa T (2013) A Microdosimetric-kinetic model for cell killing by protracted continuous irradiation including dependence on LET I: repair in cultured

mammalian cells. Radiat Res 180(6):584–594. https://doi.org/10.1667/RR13257.1

60. Bertolet A, Cortés-Giraldo MA, Carabe-Fernandez A (2021) Implementation of the microdosimetric kinetic model using analytical microdosimetry in a treatment planning system for proton therapy. Phys Med 81:69–76. https://doi.org/10.1016/j.ejmp.2020.11.024
61. Tobias CA. The repair-MISREPAIR model of cell survival. Published online June 1, 1980. Accessed 22 Nov 2024. https://escholarship.org/uc/item/8nd7k3fz
62. Kamp F, Cabal G, Mairani A, Parodi K, Wilkens JJ, Carlson DJ (2015) Fast biological modeling for voxel-based heavy ion treatment planning using the mechanistic repair-Misrepair-fixation model and nuclear fragment spectra. Int J Radiation Oncol Biol Phy 93(3):557–568. https://doi.org/10.1016/j.ijrobp.2015.07.2264
63. Cao W, Khabazian A, Yepes PP et al (2017) Linear energy transfer incorporated intensity modulated proton therapy optimization. Phys Med Biol 63(1):015013. https://doi.org/10.1088/1361-6560/aa9a2e
64. Chen M, Cao W, Yepes P et al (2023) Impact of dose calculation accuracy on inverse linear energy transfer optimization for intensity-modulated proton therapy. Prec Radiation Oncol 7(1):36–44. https://doi.org/10.1002/pro6.1179
65. Gutierrez A, Rompokos V, Li K et al (2019) The impact of proton LET/RBE modeling and robustness analysis on base-of-skull and pediatric craniopharyngioma proton plans relative to VMAT. Acta Oncol 58(12):1765–1774. https://doi.org/10.1080/0284186X.2019.1653496
66. Liu C, Patel SH, Shan J et al (2020) Robust optimization for intensity modulated proton therapy to redistribute high linear energy transfer from nearby critical organs to tumors in head and neck cancer. Int J Radiation Oncol Biol Phy 107(1):181–193. https://doi.org/10.1016/j.ijrobp.2020.01.013
67. Peeler CR, Mirkovic D, Titt U et al (2016) Clinical evidence of variable proton biological effectiveness in pediatric patients treated for ependymoma. Radiother Oncol 121(3):395–401. https://doi.org/10.1016/j.radonc.2016.11.001
68. Taleei R, Rahmanian S, Nikjoo H (2024) Modelling cellular response to ionizing radiation: mechanistic, semi-mechanistic, and phenomenological approaches—a historical perspective. Rare 202(2):143–160. https://doi.org/10.1667/RADE-24-00019.1

Chapter 6

In Vitro Analysis of Cell Cycle Exit of Cerebellar Granule Cell Progenitors

Isabelle Egan, Eve-Ellen Govek, Yin Fang, Anna Jermakowiz, Mary E. Hatten, and Nagi G. Ayad

Abstract

Despite intensive efforts, our understanding of the cell cycle exit of neuronal progenitors in the developing cerebellum is incomplete. It is important to understand this process, as disruption in cell cycle exit can be linked to cancers, such as the most common malignant pediatric brain tumor, medulloblastoma. One of the main cell types thought to give rise to medulloblastoma is cerebellar granule cell progenitors, or GCPs. GCPs give rise to cerebellar neurons, which interact with Purkinje cells, the principal output neuron of the cerebellar cortex, and they are essential for cerebellar architecture. GCPs exit the cell cycle after birth in mice and can be purified to near homogeneity, offering a robust model for studying cell cycle exit. We adapted a protocol from the Hatten laboratory for purification of GCPs, followed by plating on laminin coated plates. We then isolated GCPs at distinct time points to assess cell cycle exit parameters. To determine the transcriptional profiles of GCPs that are cycling, differentiating, or differentiated, we performed RNA-sequencing at individual time points. We performed time series analysis of all transcripts and identified times and pathways associated with cell proliferation, cell cycle exit, and neuronal differentiation. Here, we provide detailed methods for analyzing GCP cell proliferation and differentiation in vitro and how to use this information for determining cell cycle exit parameters and pathways.

Key words Cerebellum, Granule cell progenitors, Cell cycle exit, Differentiation

1 Introduction

The granule cell (GC) is the most abundant neuron type in the central nervous system [1–8]. GCs evolve from granule cell progenitors, which undergo symmetric division until exiting the cell cycle in a precise time frame, shortly after birth. Failure to exit the cell cycle at this time is associated with cerebellar abnormalities, such as cerebellar ataxia, abnormal cerebellar morphology, and medulloblastoma [8]. Because the mechanism of this cell cycle exit is not fully understood, it is critical to elucidate the factors that contribute to the timing of GCP cell cycle exit. By revealing the factors responsible for the proper timing of GCP cell cycle exit, it

Michael Goldstein (ed.), *Translational Research Methods for Pediatric Brain Tumors*, Neuromethods, vol. 226, https://doi.org/10.1007/978-1-0716-4754-7_6,

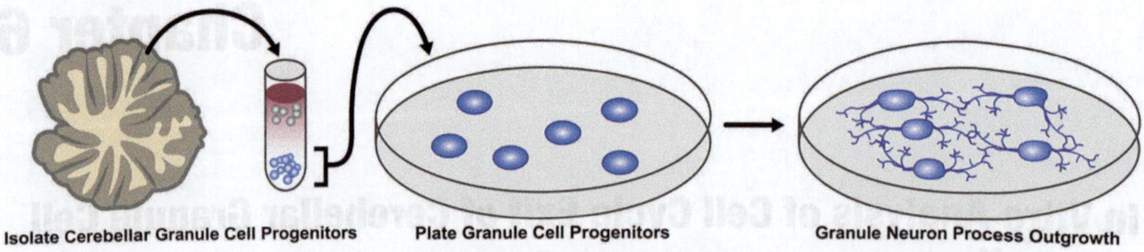

Fig. 1 Diagram of morphological changes of GCPs exiting the cell cycle to become granule cells

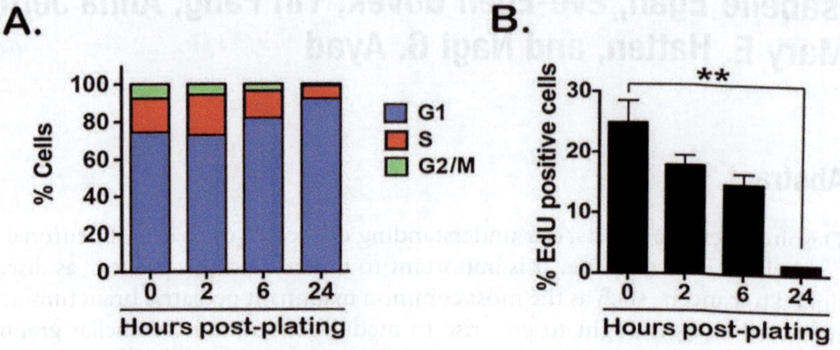

Fig. 2 PI and EdU incorporation assays. (**a**) GCPs exit the cell cycle within 24 h of plating as judged by PI-FACS analysis. FACS analysis was performed on GCPs purified from P6 mice and plated for the time intervals indicated. FlowJo software was then used to assess the percentage of cells in the G1, S, or G2/M phase. (**b**) Purified GCPs exit the cell cycle within 24 h of plating as judged by EdU incorporation studies. GCPs purified from P6 mice were processed for EdU incorporation, which was normalized to the total number of cells labeled with Hoechst staining. (From Penas et al., 2018, Nature Communications)

becomes possible to identify genetic targets for the treatment of medulloblastoma and nonneoplastic cerebellar disorders.

Although it is possible to analyze the timing of cell cycle exit of GCPs in vivo [8], transcriptional profiling of GCPs at the individual, cellular level is difficult in vivo. Therefore, we utilized an in vitro approach to analyze the transcriptome of purified GCPs (Fig. 1). The protocol we describe here yields an outstanding purity of GCPs, exceeding 98%. Thus, by achieving a high purity cell fraction in an efficient manner, our protocol offers a superior method for isolation of neural progenitors and time series analysis.

In this chapter, we report a method for analyzing the transcriptome of GCPs at various time points before and after cell cycle exit. GCPs can be obtained from the cerebellum of postnatal day 6 CD1 mice, purified, and plated. In order to analyze gene expression relative to time of cell cycle exit, it is necessary to determine the time frame of cell cycle exit experimentally. Two methods were used to accomplish this: an EdU incorporation assay and PI-FACS analysis (Fig. 2).

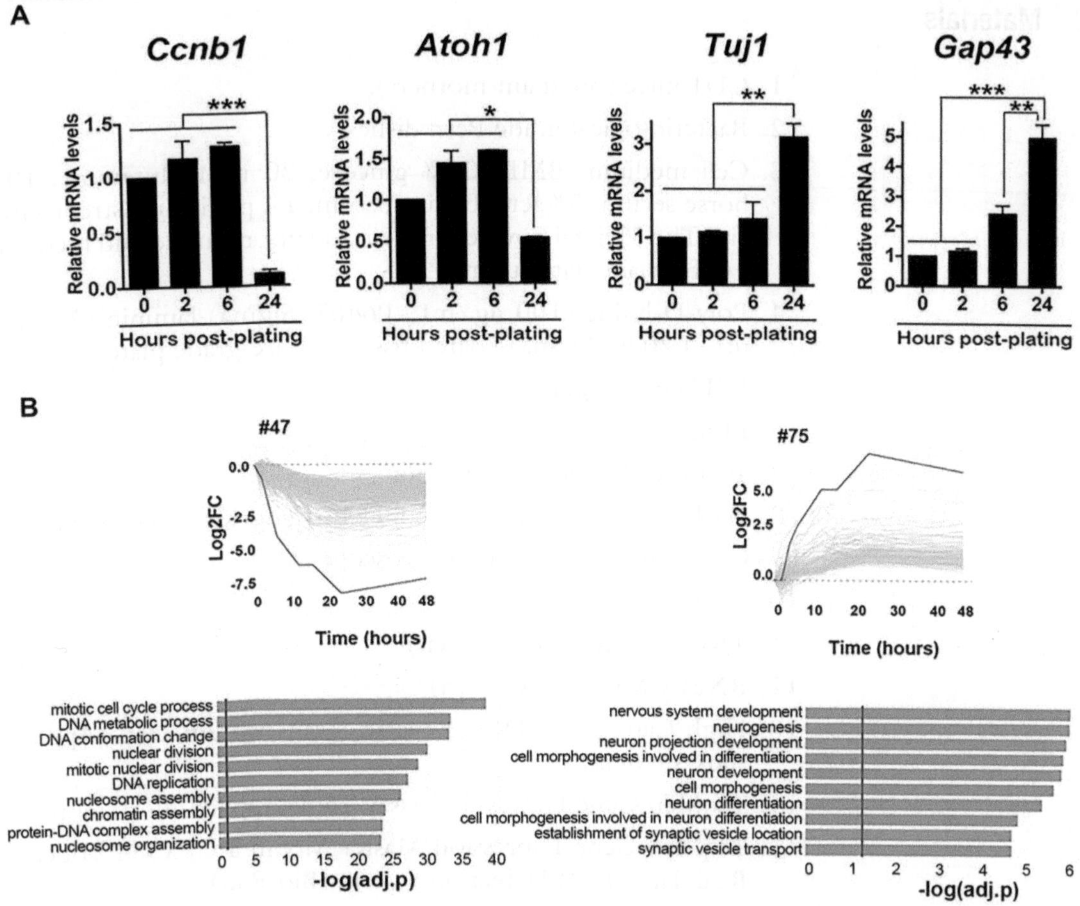

Fig. 3 Transcriptional profile changes. (**a**) Purified GCPs from P6 mice were plated and processed for RNA expression at the time intervals indicated for the proliferative markers, *Ccnb1* and *Atoh1*, and the differentiation markers, *Tuj1* and *Gap43*. qRT-PCR was performed and normalized to *Gapdh*. (**b**) Short-time series modeling of mRNAs during cell cycle exit. Plots represent mRNA expression profiles during GCP cell cycle exit. Representative clusters (#47, #75 and their related cellular processes) are shown. The red line represents the consensus for each cluster. The gray lines represent individual mRNA expression profiles. The associated biological processes as defined by DAVID41 in each cluster are shown. The results are shown as the average values of three independent experiments and are represented as the mean ± SEM. A one-way ANOVA followed by Bonferroni multiple comparison testing was performed (*$p < 0.05$, **$p < 0.01$). (From Penas et al., 2018, Nature Communications)

Using these methods, it was possible to analyze the relative levels of mRNA transcripts present at different time points (Fig. 3), and identify genes associated with cell proliferation and neuronal differentiation as well as potential therapeutic targets for cerebellar disorders.

2 Materials

1. CD1 mice (pregnant mothers).
2. Bacteriological grade Petri dishes.
3. Cell medium: BME, 0.9% glucose, 20 mM glutamine, 10% horse serum, 5% fetal bovine serum, 1% penicillin/streptomycin. The 5% fetal bovine serum is also not required and medium can be made without it.
4. Poly-D-lysine (100 μg/mL, P6407, Sigma)-laminin (20 μg/mL, L2020, Sigma)-coated tissue culture grade plates.
5. EdU (Invitrogen).
6. PBS.
7. 4% PFA/30% sucrose solution.
8. ClickiT® EdU Alexa Fluor® 594 Imaging Kit (Invitrogen).
9. Confocal laser-scanning microscope.
10. ImageJ software.
11. TRIzol reagent (Invitrogen).
12. RNeasy Mini Kit (Qiagen).
13. High-Capacity cDNA Reverse Transcription Kit (Applied Biosystems).
14. TaqMan Gene Expression Assay Tool (Applied Biosystems).
15. TaqMan Gene Expression Master Mix in a CFX384 Touch™ Real-Time PCR Detection System (Bio-Rad).
16. Thermo Scientific NanoDrop or Agilent Bioanalyzer.
17. Illumina HiSeq2000.
18. FastQC.
19. TopHat.
20. CuffDiff 2.2.
21. Pasteur pipettes.
22. Bulbs for Pasteur pipets.
23. Plastic transfer pipets: Sarstedt 86.1172.301.
24. Percoll: Sigma P4937.
25. Gibco 1× Dulbecco's Phosphate Buffered Saline (DPBS) (without Calcium Chloride and without Magnesium): 14190-144.
26. Corning 7.5% Sodium Bicarbonate Solution ($NaHCO_3$): 25-035-CI.
27. Invitrogen by Life Technologies UltraPure 0.5 M EDTA, pH 8.0: 15575-038.

28. Trypan Blue Solution 0.4%: Sigma T8154-100 mL.
29. Sodium Chloride (NaCl): Sigma S-5886.
30. Potassium Chloride (KCl): Fisher P217-500.
31. Sodium Phosphate Monobasic, Monohydrate (-$NaH_2PO_4H_2O$): Sigma S-9638.
32. Potassium Phosphate Monobasic (KH_2PO_4): Fisher P285-500.
33. Gibco Basal Medium Eagle (BME): Fisher 21-010-046.
34. 45% Glucose solution: Corning 25-037-CI.
35. Gibco 200 mM (100×) L-glutamine: 25030-081.
36. Gibco 100x Penicillin/Streptomycin solution: 15140-122, 100 ml.
37. 0.5% Phenol Red: Sigma P0290.
38. Trypsin powder: Worthington Biochemical LS003707.
39. DNase powder: Worthington Biochemical LS002139.
40. Gibco Horse serum: ThermoFisher Scientific 26050088 (note – serum should be heat-inactivated at 55 °C for 30 min, aliquoted, and stored at −20 °C).
41. N2 supplement: Gibco 17502014.
42. Gibco Sodium Pyruvate: ThermoFisher Scientific 11360070.
43. N-Acetyl-Cysteine: Sigma A9165.
44. Nylon Filters: Falcon/Corning 352340.

3 Methods

3.1 Isolation of GCPs

GCP isolation is performed essentially as described previously (Hatten 1985). GCP isolation should be performed as quickly as possible for optimal cell numbers and viability. This protocol is optimal for the isolation of GCPs from 4 to 6 P6-P7 cerebella per Percoll gradient with 10 ml of 35% Percoll and 10 ml of 60% Percoll in a 50 ml clear polystyrene tube. For less cerebella, scale solution volumes down and use a 15 ml conical tube instead with 5 ml of each percentage of Percoll. For more cerebella, set up more Percoll gradients, averaging cells from 4 to 6 cerebella per column.

3.1.1 Prepare GCP Isolation and Culturing Reagents in Advance

1. Prepare CMF-PBS, culture medium, and Percoll in advance to save time the day of the GCP isolation.
2. CMF-PBS (Tyrode's): For 500 ml—500 ml bottle of DPBS (without Ca and Mg); 2.22 ml of 45% Glucose; 270 μl of 7.5% $NaHCO_3$ solution; 250 μl of 0.5% solution of Phenol Red.

3. 4x CMF-PBS-EDTA: For 1 L - 32 g NaCl; 1.2 g KCl; 8 g glucose; 2 g $NaH_2PO_4H_2O$; 1 g KH_2PO_4; 2.13 ml of 7.5% solution $NaHCO_3$; 20 ml of 0.5 M EDTA, pH 8.0. pH solution to 7.35 and filter with 0.2 μm filter.
4. 35% Percoll (white): For 100 ml—25 ml of 4x CMF-PBS-EDTA; 35 ml Percoll; 40 ml autoclaved water. Filter with 0.2 μm filter. Filtering takes a long time.
5. 60% Percoll (blue): For 100 ml—25 ml of 4× CMF-PBS-EDTA; 60 ml Percoll; 15 ml autoclaved water; 0.3 ml of 0.4% Trypan Blue Solution. Filter with 0.2 μm filter. Filtering takes a long time.
6. GCP serum containing culture medium: Basal Medium Eagle (BME); 10% horse serum; 5% fetal bovine serum; 0.9% glucose; 1× L-glutamine; 1× Pen/Strep. GCPs are sensitive to sera. If your cultures do not look healthy, consider trying a different lot of horse serum. The 5% fetal bovine serum is also not required and medium can be made without it.
7. For some stimulation protocols, you may need to use serum-free medium for optimal stimulation. GCPs also do better with preplating in serum and short-term culturing (a few days) in serum-containing medium, but for long-term cultures, they do better with preplating in serum-containing medium followed by culturing in serum-free medium.
8. GCP serum-free culture medium: Basal Medium Eagle BME; 1× N2 supplement; 0.9% glucose; 1 mM sodium pyruvate; 1× L-glutamine; 1× Pen/Strep; 1× N-acetyl-cysteine.
9. Trypsin-DNAse solution: For 100 ml—100 mg DNAse; 1 g trypsin; 99.4 ml CMF-PBS; 0.6 ml 1 N NaOH.

Fire polish Pasteur pipettes. Will need fine and extra fine diameters. Fine is slightly constricted and extra fine about half-constricted or so after fire polishing. Important to note: If the diameter of the fire-polished Pasteur pipette is too small, you can damage the cells and lose a lot of cells, but too large and tissue and cells will be clumpy and you will lose a lot of cells.

3.1.2 The Day Before GCP Isolation

Coat wells with 100 μg/ml poly-D-lysine (P6407, Sigma) overnight at 35 °C or 37 °C. The following morning, wash the wells three times with tissue culture grade water and coat with 20 μg/ml laminin (L2020, Sigma) until you are ready to plate the GCPs.

3.1.3 The Day of GCP Isolation

1. Dissect the cerebella from P6-P7 mice in a dissection hood. Dissect cerebellum from each pup, doing one at a time, in a petri dish containing ice-cold CMF-PBS. Remove the meninges from the cerebellum and transfer the cerebellum to

a 15 ml tube with ice-cold CMF-PBS on ice. Do dissections and remove pial membranes as quickly as possible and leave in ice-cold CMF-PBS on ice until all cerebella are dissected.

2. Remove CMF-PBS with a plastic transfer pipette.
3. Add 1 ml of trypsin/DNase solution and incubate at 37 °C for 5 min.
4. Let cerebella settle and remove the trypsin/DNase solution with a transfer pipette and add 1 ml of DNase solution to cerebella.
5. Gently triturate, pipetting up and down, 10 times with a wide-bore plastic transfer pipette, being careful to avoid making bubbles.

   Tip: When triturating during this and subsequent steps, do so against the bottom of the conical tube. If you are too gentle, you will get too few cells, but if you are too harsh, you could also damage the cells.
6. Gently triturate, pipetting up and down, 10 times with a fine fire-polished Pasteur pipette, being careful to avoid making bubbles.
7. Gently triturate, pipetting up and down, 10 times with an extra-fine fire-polished Pasteur pipette, being careful to avoid making bubbles.
8. Spin cells at 2000 rpm (800 g) for 5 min at 4 °C with the brake on.

   During the centrifugation, it is best to prepare the Percoll column so that you are ready to load the column after the next cell trituration steps. Prepare one Percoll column per cell solution from 4 to 6 cerebella.
9. Place a 10 ml syringe without the plunger and with an 18 1/2G needle standing needle down in a clear polystyrene 50 ml tube in a stand. Make sure the tip of the needle is at the very bottom of the tube.
10. Add 10 ml of 35% Percoll to the 50 ml tube outside the syringe.
11. Add 10 ml of 60% Percoll (with blue dye to distinguish from 35% Percoll) inside the syringe. The 60% Percoll will exit the needle and form a layer below the 35% Percoll. Once all of the 60% Percoll has exited the syringe and needle, gently pull the syringe with the needle up and out of the gradient without disturbing the layers. Discard the syringe and needle in a sharps container. Tip: If the 60% Percoll solution is not flowing out of the syringe with needle, give an ever-so-gentle "push" (barely touch/insert) of a plunger at the very top of the syringe to get the air bubble out and start the solution flowing.

Continue with the cell trituration.

12. Remove the DNase supernatant from the cells with a plastic transfer pipette.
13. Add another 1 ml of DNase to the cell pellet and resuspend the cells with a wide-bore plastic transfer pipette, being careful to avoid making bubbles.
14. Gently triturate, pipetting up and down, 10 times with an extra fine fire-polished Pasteur pipette, being careful to avoid making bubbles.
15. Make sure the cell suspension is homogeneous and there are no clumps of cells. Optionally, you can also pass the cell suspension through a 40 μM nylon mesh filter to ensure you have single cells.
16. Place the tip of the Pasteur pipette against the inside wall of the conical tube near the Percoll gradient and gently release so that the cell solution flows down gently against the tube wall, and you layer the cells on top of the Percoll gradient. Do this step very slowly and gently.
17. Spin the Percoll gradient at 3000 rpm (2000 g) for 10 min at 4 °C with the brake off.
18. Remove the column(s) from the centrifuge gently so as not to disturb the gradient layers.
19. The top layer of cells contains the large cell fraction—the glial cells, Purkinje cells, and fibroblasts. The middle layer of cells, at the interface between the 35% Percoll and 60% Percoll layers, contains the GCPs in the small cell fraction. Remove and discard the top, pink layer of cells with a wide-bore plastic transfer pipette, being careful not to disturb the layers below, and discard if you are not interested in these cell types. Do not remove too much of the lower 35% Percoll layer while you are removing the top layer.
20. Remove the granule cell layer, the thin, white floating layer between the 35% Percoll and 60% Percoll by "vacuuming" them gently with a clean, wide-bore plastic transfer pipette, being careful not to disturb the Percoll layers. Transfer them into another 50 ml tube with about 40 ml of ice-cold CMF-PBS on ice. You can take up a little of the blue 60% Percoll layer and take most of the 35% Percoll layer until you get all of the cells, but avoid the top, pink layer with the large cell fraction and taking up too much of the Percoll, which can ultimately be toxic to the cells.
21. Fill the 50 ml tube with the small cell GCP fraction to 50 ml with ice-cold CMF-PBS and invert a couple of times very gently.
22. Spin at 2000 rpm (800 g) for 5 minutes at 4 °C with the brake on.

After this centrifugation, move the tube with cells to a tissue culture hood.

23. Aspirate off the supernatant gently so as not to aspirate the tiny pellet, flick the tube gently to loosen/break up the pellet, and resuspend the cells gently in 200 ul of GCP medium. If the cells are clumpy/viscous, you can use a small amount of DNase (about 100 μl) to resuspend the gloopy pellet, but the DNase can be toxic, so only do this if necessary.
24. Transfer the cells to a 60 mm bacteriological grade Petri dish with 3 ml of GCP culture medium (can combine 2 gradients of cells for the same conditions into one dish) and preplate at 35 °C for 20–30 minutes. Fibroblasts will adhere to these Petri dishes.
25. Tap the edge of the plate gently against the side of the hood grill. Tap gently to avoid dislodging large cells, but check the dish under the microscope to also make sure you are not losing too many GCPs, the small cells. Large cells adhering to the bottom are fibroblasts. Small cells are the GCPs.

    Tip: If using serum free medium, the cells tend to stick down more and you may need to gently wash the cells/ medium over the bottom of the dish several times with the transfer pipette, but if cells are in serum, tapping may be enough for this and subsequent preplating steps.
26. Transfer the cells/medium to a regular 60 mm tissue culture dish and incubate at 35 °C for 30 minutes to 1 hour to remove glia and Purkinje cells.
27. Tap the edge of the plate gently against the side of the hood grill.
28. Transfer cells/medium to another regular 60 mm tissue culture dish and incubate at 35 °C for 30 minutes to 1 h to remove glia and Purkinje cells.
29. Tap edge of plate gently against the side of the hood grill.
30. Transfer the cells/medium to another regular 60 mm tissue culture dish and incubate at 35 °C for 30 minutes to 1 h to remove glia and Purkinje cells.
31. Tap the edge of plate gently against the side of the hood grill.
32. Transfer the GCPs cells to a conical tube, count (GCPs are the small cells), and plate.
33. Plate about 300,000 GCPs in 300 μl of medium in a 16-well slide or 96-well plate, and scale up for larger wells. (Can go down to 100,000 for a sparse culture, but not lower. They do best when they are more densely plated, and you will need higher density to obtain adequate RNA.)
34. Plate GCPs in poly-D-lysine and laminin-coated wells for various indicated lengths of time.
35. Maintain the cells in a 5% CO_2, 35 °C incubator.

3.2 Determination of Cell Cycle Exit Using PI-FACS Analysis and/or EdU Incorporation Assay

PI-FACS Analysis

1. At various lengths of time, harvest GCPs and fix with 4% PFA for 30 minutes. Wash away PFA, and then, add propidium iodide and RNAse. Perform FACS analysis.
2. Use FlowJo software to assess the percentages of cells in the G1, S, or G2/M phase. Most of the cells are in G1 phase with a lower proportion in S phase and M phase.

EdU Incorporation Assay

1. Treat plated GCPs with EdU (20 μM) for 2 hours.
2. Wash cells with PBS. Replate in poly-D-lysine/laminin-coated dishes for 2–3 hours.
3. Fix cells with 4% PFA/30% sucrose.
4. Use ClickiT® EdU Alexa Fluor® 594 Imaging Kit (Invitrogen) to permeabilize and stain cells.
5. Acquire z-stacks of cells using a confocal laser-scanning microscope and analyze images using ImageJ software.

3.3 Isolation of mRNA from GCPs and qRT-PCR

Conduct procedure at selected time points: 0, 2, 4, 6, 12, 24, and 48 h post-plating.

1. Lyse cells in 1 mL of TRIzol reagent (Invitrogen).
2. Purify RNA using RNeasy Mini Kit (Qiagen).
3. Reverse-transcribe the RNA using High-Capacity cDNA Reverse Transcription Kit (Applied Biosystems).
4. Use TaqMan Gene Expression Assay Tool (Applied Biosystems) to design TaqMan probes.
5. Conduct PCR using TaqMan Gene Expression Master Mix in a CFX384 Touch™ Real-Time PCR Detection System (Bio-Rad).
6. Use the computed tomography comparative method to estimate fold change in gene expression. Normalize to the *Gapdh* cycle threshold (CT) values and relative to control samples.

3.4 RNA Sequencing of GCPs Exiting the Cell Cycle

1. Test RNA quality using Thermo Scientific NanoDrop or Agilent Bioanalyzer. Confirm RNA integrity number (RIN) numbers >8.5.
2. Send extracted RNA for analysis.
3. Conduct RNA sequencing on Illumina HiSeq2000 with three samples per lane.
4. Filter and trim RNA-seq reads with FastQC.
5. Use TopHat to align remaining RNA-seq reads to the Ensembl mouse genome (v.87).

6. Perform differential expression analysis with CuffDiff 2.2. Use the "classic-fpkm" parameter for normalization and the "pooled" parameter for dispersion.
7. Perform functional annotation analysis using DAVID against the Level 5 Gene Ontology Biological Processes terms (GOTERM_BP_5). To adjust for false discovery, consider only terms with a Benjamin-Hochberg adjusted *p*-value of 0.05.

4 Notes

We have found that it is essential to isolate cerebella and remove GCPs from Percoll as quickly as possible. Viability of GCPs should be close to 98%. It is possible to use only horse serum, and no fetal bovine serum, in the cell medium or to use serum-free medium for GCP cultures. It is possible to reduce preplating times in order to optimize the timing of cell cycle exit in vitro.

Acknowledgments

We thank all members of the Ayad laboratory for helpful discussions. This work was supported by NS118023 and BellRinger at Georgetown Lombardi Comprehensive Cancer Center.

Conflict of Interest The authors have no conflicts of interest to declare that are relevant to the content of this chapter.

References

1. Penas C, Govek EE, Fang Y, Ramachandran V, Daniel M, Wang W et al (2015) Casein kinase 1δ is an APC/C(Cdh1) substrate that regulates cerebellar granule cell neurogenesis. Cell Rep 11(2):249–260
2. Gao WQ, Heintz N, Hatten ME (1991) Cerebellar granule cell neurogenesis is regulated by cell-cell interactions in vitro. Neuron 6(5): 705–715
3. Roussel MF, Hatten ME (2011) [cited 2024 Oct 7] Cerebellum. In: Current topics in developmental biology [internet]. Elsevier, pp 235–282. Available from: https://linkinghub.elsevier.com/retrieve/pii/B9780123809162000085
4. Hatten ME, Roussel MF (2011 Mar) Development and cancer of the cerebellum. Trends Neurosci 34(3):134–142
5. Lee HY, Greene LA, Mason CA, Manzini MC (2009) Isolation and culture of post-natal mouse cerebellar granule neuron progenitor cells and neurons. J Vis Exp 16(23):990
6. Carletti B, Rossi F (2008 Feb) Neurogenesis in the cerebellum. Neuroscientist 14(1):91–100
7. Hatten ME (1985 Feb) Neuronal regulation of Astroglial morphology and proliferation in vitro. J Cell Biol 100(2):384–396
8. Penas C, Maloof ME, Stathias V, Long J, Tan SK, Mier J et al (2019) Time series modeling of cell cycle exit identifies Brd4 dependent regulation of cerebellar neurogenesis. Nat Commun 10(1):3028

Chapter 7

Using the RCAS/tv-a System and Cre/loxP Recombination to Establish a Primary Mouse Model of Diffuse Midline Glioma

Spencer M. Maingi and Zachary J. Reitman

Abstract

Diffuse midline gliomas (DMG) are aggressive brain tumors with few treatment options that affect children and young adults. While recent genomic profiling of brain tumors has significantly enhanced our understanding of tumor biology, this knowledge has not yet translated to improved outcomes for patients with DMG. Given this lack of progress, there is a pressing need for innovative modeling approaches that faithfully recapitulate DMG. Here we discuss how a fusion of Cre/loxP genetic engineering and the RCAS/tv-a retroviral delivery system can be used to induce DMG tumors in mice. Notably, this platform allows tumors to occur alongside murine neurodevelopment while also preserving the immune system and genotypes of non-neoplastic cells, allowing for faithful study of tumor development and tumor-immune interactions, respectively.

Key words Diffuse midline glioma, Brain tumors, Primary mouse model, RCAS/tv-a system, Cre/loxP genetic engineering

1 Introduction

Although high grade pediatric brain tumors occur rarely, they represent an urgent clinical need. One subtype is diffuse midline gliomas (DMG), which are found in critical neuroanatomical regions like the brainstem and thalamus and are frequently driven by activating mutations in histone 3 genes such as *H3F3A*, and by mutations that inactivate the tumor suppressor *TP53* pathway [1–4]. Between 200 and 400 new DMG diagnoses are made each year in the United States [5], but each is life-limiting; less than 10% of patients will survive beyond 2 years [6]. Radiation therapy has been adopted as the standard of care at diagnosis and in some cases at disease recurrence, but its use only minimally extends lifespan. Due to the tumor's location and infiltrative nature, complete surgical

Michael Goldstein (ed.), *Translational Research Methods for Pediatric Brain Tumors*, Neuromethods, vol. 226,
https://doi.org/10.1007/978-1-0716-4754-7_7,

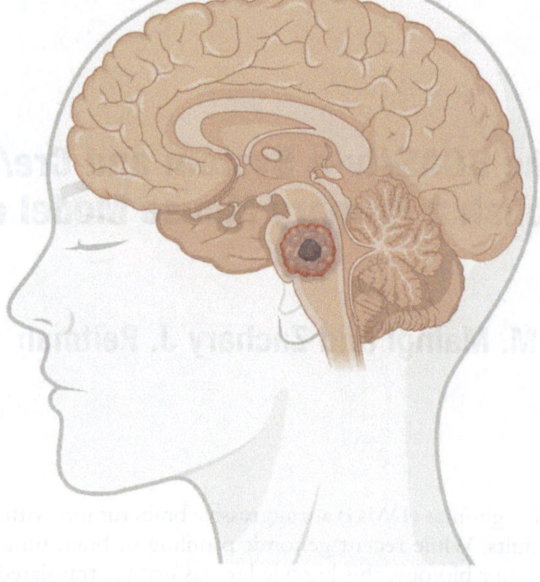

Fig. 1 Schematic of a DMG tumor located in the pons. Image created with BioRender.com

resection is impossible (*see* Fig. 1). As such, novel therapeutic approaches are needed to improve DMG's dismal prognosis.

Like other brain tumors, DMGs can be modeled using patient-derived cell lines and patient-derived xenografts. These approaches are particularly valuable for testing novel therapies in an orthotopic setting [7]. However, tumor engraftment does not occur in the context of neurodevelopment and is limited to immunocompromised hosts lacking an intact adaptive immune system, a roadblock to investigating immunotherapies. Passaging cells over long periods may also select for phenotypes that do not represent the biology of tumors found in patients [8].

Several groups have developed primary mouse models using genetic engineering, Cre/loxP recombination, and/or intrauterine electroporation (IUE) as alternatives to engraftment-based tumor modeling. Larson et al. bred mice that constitutively expressed Cre in *Nestin+* neural stem cells with those that have knock-in *H3.3K27M* and *PDGFRA* alleles plus *Tp53* loss to achieve lineage-specific tumorigenesis [9]. IUE has also been used to induce DMG formation through the delivery of oncogenic plasmids to the mouse brainstem [10, 11]. Notably, both constitutive and IUE approaches occur in the context of the developing brain of an immune-competent host.

Other primary DMG models use the replication-competent avian-like sarcoma virus/tumor virus receptor-A (RCAS/tv-a) system. RCAS viruses transfect and replicate only within cells that

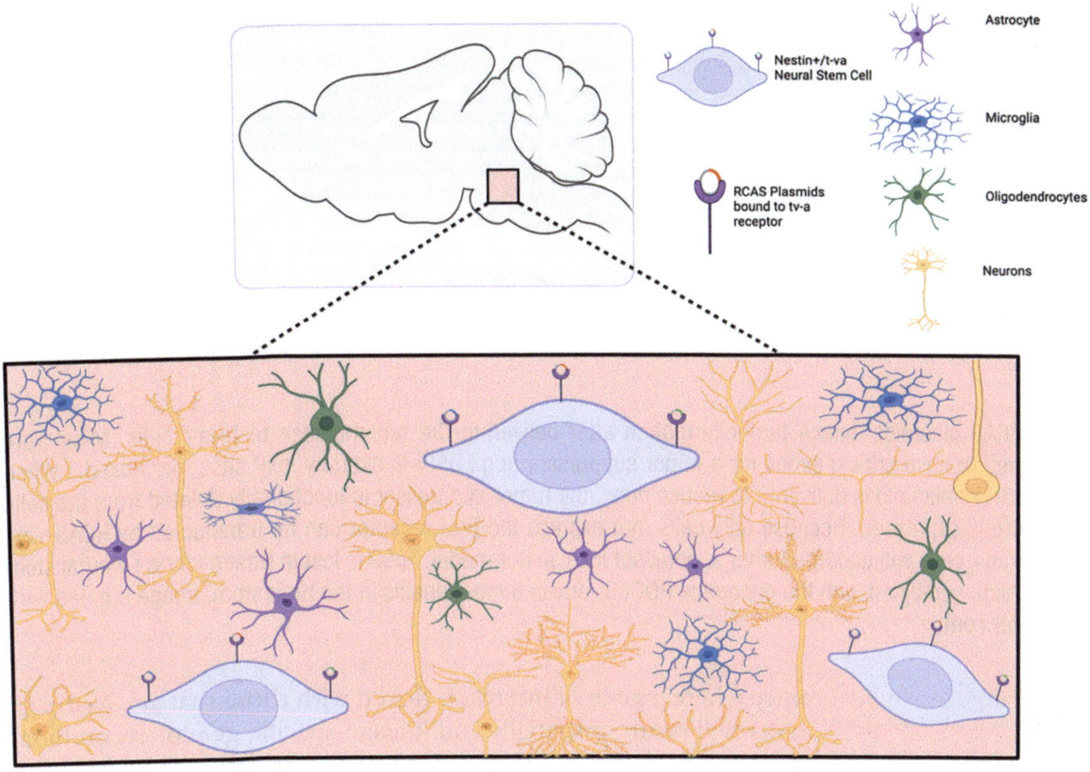

Fig. 2 Transfection of *Nestin+* neural stem cells (shown in purple) by RCAS viruses. Mice were engineered to express the tv-a receptor under the control of the Nestin promoter, which allows for lineage-specific delivery of genetic information. Other cell types in the brain will not be transfected by the RCAS retroviruses because they lack the tv-a receptor. Image created with BioRender.com

express the tv-a receptor, allowing for lineage-specific delivery of oncogenic payloads like PDGFB, H3.3K27M, and ACVR1 R206H [12, 13]. This dependency can be leveraged by engineering mammalian cell populations to express the tv-a receptor under the control of a specific promoter. Indeed, previous work has shown that *Olig2*+/tv-a+, *Pax3*+/tv-a+, and *Nestin*+/tv-a+ can each serve as the cells of origin for inducing gliomas in mice with the RCAS/tv-a system [14, 15]. If these cell populations are in specific niches of the brain (e.g., *Pax3*+ and *Nestin*+ cells, which are confined to the neonatal murine pons), RCAS exogenous gene delivery will be spatially restricted too (*see* Fig. 2). Like the IUE and constitutive model described above, these RCAS-induced DMGs also arise in the appropriate neurodevelopmental setting with the immune system of the host intact.

RCAS/tv-a oncogene delivery is often accompanied by Cre/loxP genetic engineering and a mouse breeding scheme to enhance tumorigenesis and tissue specificity. When Cre recombinase is introduced, mice with a gene flanked by loxP sites ("floxed") undergo removal ("floxing") of that gene (*see* Fig. 3). Crossing

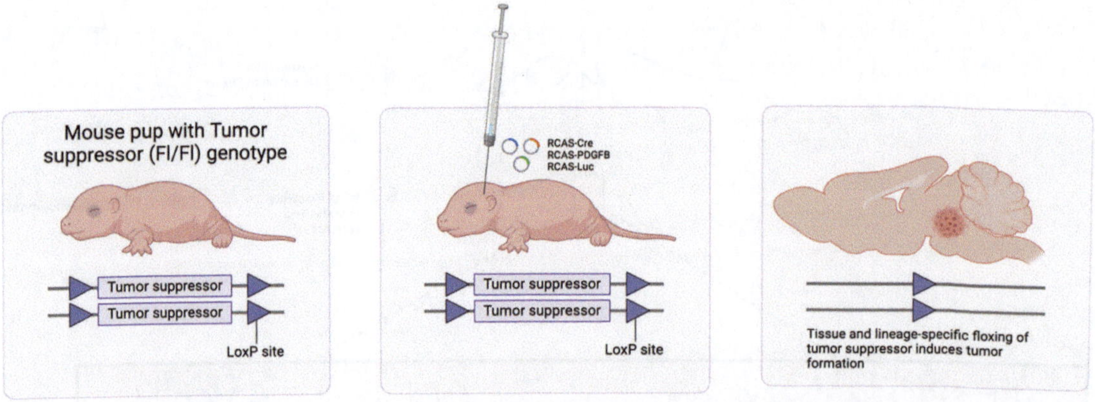

Fig. 3 RCAS plasmids induce tumor formation after delivering Cre recombinase to tv-a+ cells. Mice were engineered to have critical exons for a tumor suppressor (e.g., p53) flanked by loxP sites, or "floxed." When the RCAS-*Cre* plasmid is delivered to mouse pups, the tumor suppressor is functionally deleted from the cells where Cre is expressed. Because only cells that express the tv-a receptor can be transfected by RCAS-*Cre*, only Nestin+ cells will undergo tumor suppressor loss. In our model, Tp53 is lost in Nestin+/tv-a+ neural stem cells, which, combined with the oncogene PDGFB, drives tumorigenesis in the brainstem. Image created with BioRender.com

mice where a gene of interest is floxed with those that are *Nestin+*/tv-a+ will result in spatially and lineage-specific genetic recombination upon delivery of an RCAS-*Cre* retrovirus. Notably, this approach maintains the genotype of all non-neoplastic cells in the brain.

Building on the work of Oren Becher and others, we have developed a similar system for modeling DMGs using the RCAS/tv-a system and Cre/loxP genetic engineering [16]. Using RCAS-*PDGFB*, RCAS-*Cre*, and RCAS-*Luc* plasmids, we generate tumors in neonatal mice that lack a tumor suppressor (e.g., *Tp53* or *Pten*), bioluminesce prior to the onset of neurological symptoms, and are lineage and spatially restricted to a population of *Nestin+* neural stem cells in the brainstem. This approach has allowed our lab to investigate therapeutic targets like the serine/threonine kinase ATM [17, 18] and to gain deeper insights into DMG biological dependencies (*see* Fig. 4).

Like any other model, this approach has limitations. Although tumors develop in the proper anatomical and neurodevelopmental context, RCAS payload delivery is limited to a 2.5 kb payload size [19]. Previous work using CRISPR-Cas9 and the RCAS/tv-a system successfully generated high-grade gliomas in *Nestin+* and *GFAP+* cells [20], an approach that could be extended to DMGs in the future. In addition, tumors develop over time even amongst neonates from the same litter, which can make planning experiments logistically challenging. It's important to recognize that using orthogonal models is crucial for a complex and deadly disease like DMG because it maximizes the likelihood that preclinical

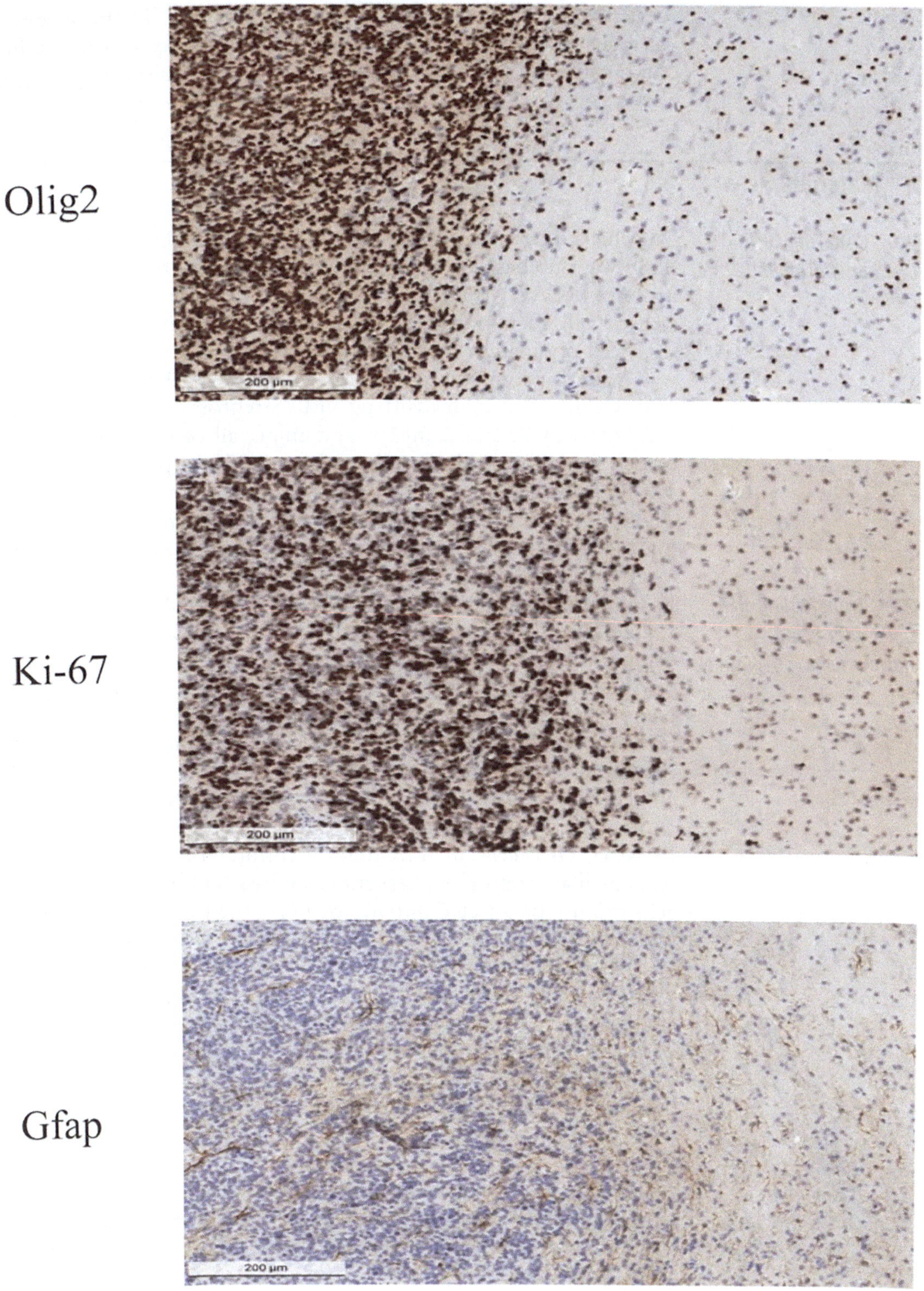

Fig. 4 Representative IHC staining for canonical DMG marker *Olig2*, proliferative marker *Ki-67*, and glioma marker *Gfap* from a murine diffuse thalamic glioma

results will one day translate to patient outcomes. In this chapter, we focus on generating DMG tumors in mice using the unique combination of Cre/loxP and RCAS/tv-a technologies described above.

2 Materials

2.1 Culturing DF1 Chicken Fibroblast Cells

The chicken embryonic fibroblast cell line DF1/UMNSAH/CRL-3586 was purchased from ATCC. These cells will be transfected to produce RCAS viruses and injected into the brain. The RCAS viruses will then transfect the engineered *Nestin*+/tv-a+ neural stem cell population to form brain tumors. DF1 culturing media was made by combining DMEM with 10% Fetal Bovine Serum, 100 U/mL Penicillin, 100 μg/mL Streptomycin, 2.5 μg/mL amphotericin B, and 2 mM L-glutamine, all of which were purchased from ThermoFisher. Phenol red 0.25% trypsin/EDTA and 1X Phosphate Buffered Saline (PBS) were also obtained from ThermoFisher. *T75* (Catalog # 10861-646) and *T25 Flasks* (Catalog # 10861-642) with vent caps were ordered from VWR.

2.2 Transfecting DF1 Cells with RCAS Retroviruses

RCAS plasmids (RCAS-*Luc*, RCAS-*GFP*, RCAS-*Cre*, RCAS-*PDGFB*) were obtained from Oren Becher's lab [21] and amplified using Zymo Research Prep Kits according to the manufacturer's instructions. The plasmids' concentrations were determined using a NanoDrop spectrophotometer, and they were sequenced to confirm the absence of mutations in protein coding regions. Each plasmid will be combined with DMEM and X-tremeGENE 9 (purchased from SigmaAldrich, Product # 6365779001; can also use another lipid-based transfection reagent) in an Eppendorf tube to transfect DF1 cells in *T25 flasks* according to the transfection reagent manufacturer's instructions. While GFP-transfected DF1 cells will not be injected into mouse pups, these cells are an important control to verify successful RCAS retrovirus transfection.

Extra precautions should be taken when working with RCAS viruses or cells transfected by RCAS viruses. In addition to normal cell culture PPE, oversleeves (VWR Catalog # 76304-468) should be worn while culturing transfected DF1 cells. A 10% sodium hypochlorite solution should also be prepared to decontaminate any serologicals, pipettes, and flasks used to culture transfected cells for at least 24 hours before disposal.

2.3 Passaging Transfected DF1 Cells

DF1 culturing media, 0.25% trypsin/EDTA, and 1X PBS are used while passaging RCAS-transfected cells in *T25 flasks*. As above, extra precautions should be taken with oversleeves and a 10% hypochlorite solution.

2.4 Harvesting Transfected DF1 Cells

Alongside oversleeves and the 10% hypochlorite solution, DF1 culturing media, 0.25% Trypsin/EDTA, and 1× PBS are used to harvest RCAS-transfected cells for injections. We use the Cellometer Auto 2000 Counter along with Cellometer PD100 slides to count transfected cells (Nexcelom Catalog # CHT4-PD100-002). A 50 mL conical should be used to centrifuge the cells so they can be resuspended at a ratio of 100,000 cells/μL.

2.5 Injecting Neonates with RCAS Retroviruses to Induce DMGs

After harvesting, transfected DF1 cells should be placed on ice in a secondary container (e.g., a cooler). This container should also have a layer of aluminum foil on top to place the mouse neonates and a 15 mL conical tube filled with DI water (for washing the Hamilton syringe). To inject the mice with the transfected DF1 cells, we use a 10 μL Hamilton syringe with a removable needle (Model 1801, VWR Catalog # 60375-806).

3 Methods

3.1 Culturing DF1 Chicken Fibroblast Cells

1. Follow recommended procedures for thawing and initiating cultures upon receipt of the DF1 vial (see **Notes 4.1**). Culture cells at 39 °C and 5% CO_2 in a *T75 flask*.
2. When DF1 cells are near/at confluency, remove spent media, wash with PBS, and split with 1 mL 0.25% Trypsin/EDTA for 5 min at 39 °C and 5% CO_2.
3. Following trypsinization, resuspend cells in an additional 5 mL of DF1 Media. Take 1 mL of cells from the old flask and add to a new *T75 flask* with 15 mL of fresh media (1:6 split).
   (a) Split the cells three times per week.

3.2 Transfecting DF1 Cells with RCAS Retroviruses

The Day Prior to Transfection

1. Ensure DF1 cells are 100% confluent prior to beginning.
2. Warm DF1 media, 0.25% trypsin/EDTA, and 1× PBS in bead bath for ~30 min.
3. Gather reagents, prepare BSC workstation, and label flasks appropriately.
4. Split the cells at a 1:12 ratio into *T25 flasks*, allocating a new flask for each RCAS virus. This splitting ratio is to ensure that DF1 cells are ~30% confluent when transfected.

The Day of Transfection

1. Warm DMEM to RT in bead bath for ~30 min. While waiting, use predetermined RCAS plasmid concentrations to calculate the volume necessary to obtain 2.5 μg.

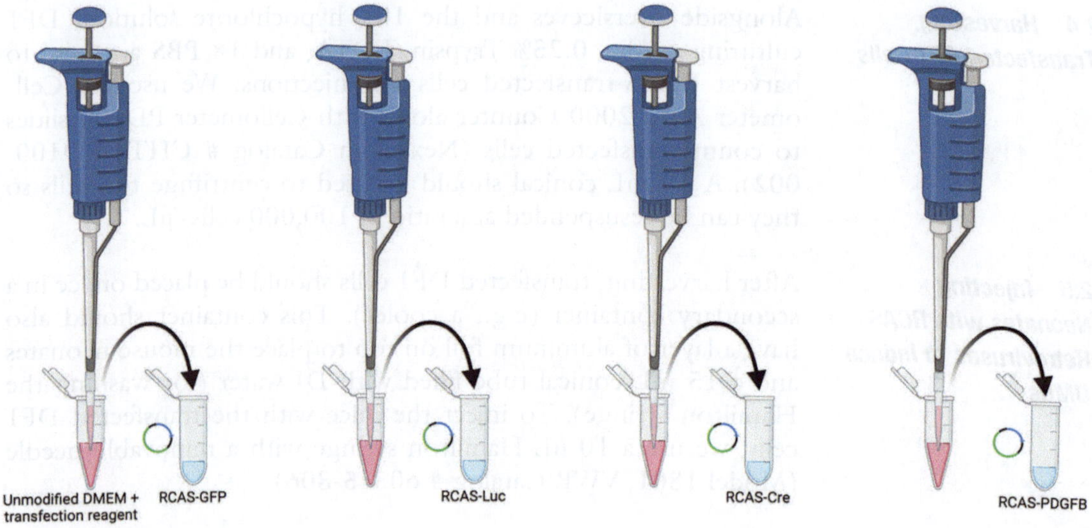

Fig. 5 Representative image showing transfection. Each aliquot of the solution containing DMEM + transfection reagent should be added to the respective RCAS plasmid aliquot. Image created with BioRender.com

2. Gather reagents, prepare BSC workstation, and label flasks and tubes appropriately.
3. Add 242.5 μL of DMEM (not DF1 media) to 1.7 mL Eppendorf tubes (one for each RCAS plasmid, Fig. 5).
4. Add 7.5 μL of X-tremeGENE 9 directly into each aliquot of DMEM to minimize contact with plastic.
5. Allow DMEM/X-tremeGENE9 transfection solution to incubate at RT for 5 min.
6. During the incubation period, aliquot 2.5 μg of each RCAS plasmid into separate 1.7 mL Eppendorf tubes.
7. When the 5 min incubation period has ended, add entirety of DMEM/X-tremeGENE 9 transfection solution to the Eppendorf containing each plasmid. Flick for ~10 s to ensure solution is evenly mixed and incubate at RT for 20 min.
8. Add each plasmid-DMEM-X-tremeGENE 9 solution dropwise to appropriate *T25 flask* and incubate at 39 °C at 5% CO_2 (see **Notes 4.2**).

3.3 Passaging Transfected DF1 Cells

1. Warm DF1 media, 0.25% Trypsin/EDTA, and 1× PBS in bead bath for ~30 min.
2. Check confluence of each *T25 flask* and determine the appropriate ratio for passaging.
3. Gather reagents, prepare BSC workstation, and label flasks appropriately.

4. Aspirate old DF1 media from flasks and wash each with 1 mL PBS.
5. Remove PBS, add 0.7 mL of 0.25% trypsin/EDTA, and incubate at 39 °C and 5% CO_2 for 5 min (see **Notes 4.3**).
6. During the incubation period, add DF1 media into new flasks (5 mL for a 1:6 split).
   (a) Note that this volume will likely change depending on the splitting ratio.
7. Using brightfield microscopy, ensure cells are detached from the bottom of *T25 flask* after trypsinizing, then neutralize the trypsin with a volume of DF1 media appropriate for the splitting ratio (5.3 mL for a 1:6 ratio).
   (a) Triturate with a serological pipette to ensure trypsin is fully neutralized and mixed with new media.
8. Add the necessary volume of resuspended cells (1 mL for 1:6 split) into a new flask filled with DF1 media.
9. Ensure cells are present in the flask, then incubate at 39 °C and 5% CO_2 until next period of passaging.
   (a) Transfected DF1 cells usually need to be passaged three times a week. Ensure these cells are not oversplit because flasks with few cells may have decreased viability.

3.4 Harvesting Transfected DF1 Cells

1. Prior to harvesting, ensure that transfected DF1 cells are near or at 100% confluency. We recommend only using transfected DF1 cells that are between passages 3–20.
   (a) Since DF1-RCAS-*GFP* cells will not be injected into mice, harvesting this flask is not required.
2. Passage cells according to the protocol above using a 1:6 split. The new flasks will continue the line for future use while the old flasks will be harvested and used for injections.
   (a) Incubate the newly passaged cells as normal until they reach confluency.
3. Using the old flasks, determine cell counts for each transfected flask and add cells at the same ratio (1:1:1 for RCAS-*Cre*, RCAS-*PDGFB*, and RCAS-*Luc* plasmids) into a 50 mL conical tube.
   (a) To find the correct volume of each transfected cell line to add, find the total number of cells in the flask with the smallest cell count. Then, divide this value by each of the remaining flasks' cells/volume ratio (see table in **Notes 4.4** for sample calculation).

4. Centrifuge the 50 mL conical down at 300×g for 5 min to form a cell pellet. If a pellet is not present, spin the cells down again or wait to harvest at a higher confluency.
5. Discard the supernatant and prepare to resuspend the pellet in DF1 media to a final concentration of 100,000 cells per μL.
   (a) To calculate the volume of DF1 media for resuspension, find the sum of all cells added to the 50 mL conical (i.e., cell counts multiplied by volume added to conical) and divide by 100,000.
6. Keep transfected DF1 cells on ice for up to 1 hour while preparing for intracranial injections.

3.5 Injecting Neonates with RCAS Retroviruses to Induce DMGs

1. Cover the portion of the ice-filled container with a layer of aluminum foil and rinse the microliter syringe with DI water.
2. To anesthetize mouse pups, place on a covered container with ice and wait until no movement is observed.
   (a) Note that mouse pups should be 3–5 days postnatal to ensure they can be sedated with hypothermia.
3. Draw up 1 μL of resuspended DF1 cells in microliter syringe. Aim needle between the pup's ears at ~15° angle (see **Notes 4.5**) and slowly inject 2 mm below skin (*see* Fig. 6). Repeat with the remaining pups.
4. After injections, warm pups and return them to the cages.
5. Begin monitoring for tumor development via neurological checks and/or bioluminescent imaging (BLI) 4 weeks post-injection. For more details on BLI procedure, consult Weidenhammer et al., STAR Protocols, 2023 [22].

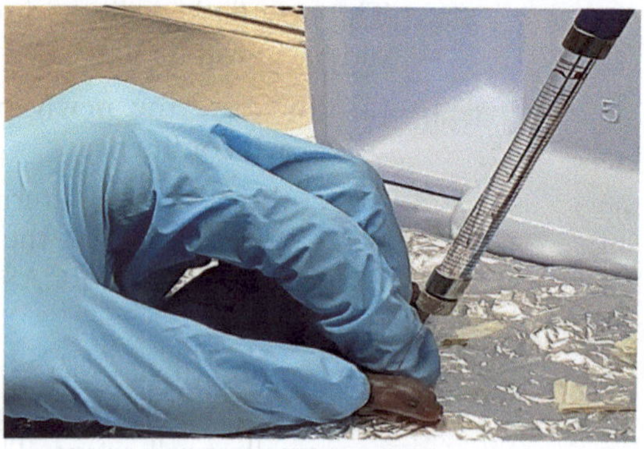

Fig. 6 DF1 cell injection into a postnatal mouse pup anesthetized via hypothermia. Needle should be aimed between the ears so that tumor-producing RCAS viruses are delivered to the *Nestin+*/tv-a+ population in the developing pons

4 Notes

1. Following initial receipt and thawing of DF1 cells, we recommend growing them to 100% confluency, then freezing down aliquots in liquid nitrogen to have a stock for future use. This can be done by trypsinizing the cells as normal, spinning them down in a 50 mL conical at 300xg for 5 min, removing the supernatant, and resuspending them in a mixture of DF1 media and 10% DMSO. The cells can then be aliquoted into cryovials and placed into a −80 °C freezer, and after 24–48 h, they can be moved to liquid nitrogen.
2. After a period of 24 h, brightfield and fluorescence microscopy can be used to verify GFP expression in the appropriate DF1 flask.
3. Consider trypsinizing only two or three flasks at a time if new to cell culture; this will minimize errors due to differences in trypsin exposure time.
4. When finished with a pipette or serological, it is typical to directly place it in a beaker of 10% sodium hypochlorite solution to decontaminate for 24 h. Similarly, one can also aliquot a small volume of this solution (enough to cover the bottom) to decontaminate old cell culture flasks for 24 h; the media in a flask should turn purple, then become clear. We recommend changing the sodium hypochlorite solution every few days.
5. It is recommended to periodically check for fluorescence in DF1-RCAS-*Luc* and DF1-RCAS-*PDGFB* flasks to ensure no contamination between cultures has occurred.
6. In Table 1, the cells/mL were obtained using a cell counter. The total number of cells for each flask is equal to the volume remaining in the flask (5 mL; recall that 1 mL was removed to continue culturing the line) multiplied by the cell count. All the cells in the flask with the smallest cell count (in the example below, RCAS-*PDGFB* with $1.0*10^6$ cells) will be added to the 50 mL conical. To ensure that each RCAS retrovirus is present in equal proportions in the 50 mL conical, we divide the total

Table 1
Sample calculations for transfected DF1 cell harvesting

Plasmid	Volume in old T25 flask (mL)	Cells/mL	Total number of cells	Volume to add to 50 mL conical (mL)
RCAS-*Luc*	5	$3.0*10^5$	$1.5*10^6$	3.3
RCAS-*Cre*		$5.2*10^5$	$2.6*10^6$	1.9
RCAS-*PDGFB*		$2.0*10^5$	$1.0*10^6$	5.0

number of cells in the flask with the smallest cell count (RCAS-*PDGFB*, $1.0*10^6$ cells) by the cell counts for the other two flasks (RCAS-*Luc*: $3.0*10^5$ cells/mL; RCAS-*Cre*: $5.2*10^5$ cells/mL) to obtain the volume from each flask to add to the 50 mL conical (3.3 mL for RCAS-*Luc*, 1.9 mL for RCAS-*Cre*).

7. Injection technique may vary slightly depending on the operator, but most use a similar approach to Fig. 6, where one hand stabilizes the neonate's head while the other slowly injects.
8. We recommend having a few extra microliters in the Hamilton syringe than the number of pups on ice at a given moment. For example, if five pups are to be injected, 7–8 μL of transfected DF1 cells should be drawn up. This helps minimize the chance that air bubbles are injected into the mouse brain.
9. Reflux of the injected cells can be limited by removing the needle slowly.
10. To prevent cross-contamination, we recommend using different microliter syringes for each unique combination of RCAS retroviruses.

Acknowledgments

We thank Dr. Oren Becher for guidance using the RCAS/tv-a system to induce DMGs. We thank Dr. David G. Kirsch for his expertise in using the Cre/loxP system and for the design of radiation biology studies. We thank Dr. Suzanne Baker for sharing the loxP-STOP-loxP-H3f3a-K27M-Tag allele.

Conflict of Interest The authors have no conflicts of interest to declare that are relevant to the content of this chapter.

References

1. Louis DN, Perry A, Wesseling P et al (2021) The 2021 WHO classification of tumors of the central nervous system: a summary. Neuro Oncol 23:1231–1251. https://doi.org/10.1093/neuonc/noab106
2. Price G, Bouras A, Hambardzumyan D, Hadjipanayis CG (2021) Current knowledge on the immune microenvironment and emerging immunotherapies in diffuse midline glioma. EBioMedicine 69:103453. https://doi.org/10.1016/j.ebiom.2021.103453
3. Mackay A, Burford A, Carvalho D et al (2017) Integrated molecular meta-analysis of 1,000 pediatric high-grade and diffuse intrinsic pontine glioma. Cancer Cell 32:520–537.e5. https://doi.org/10.1016/j.ccell.2017.08.017
4. Lewis PW, Müller MM, Koletsky MS et al (2013) Inhibition of PRC2 activity by a gain-of-function H3 mutation found in pediatric glioblastoma. Science 340:857–861. https://doi.org/10.1126/science.1232245
5. Ostrom QT, Price M, Ryan K et al (2022) CBTRUS statistical report: Pediatric Brain Tumor Foundation childhood and adolescent primary brain and other central nervous system tumors diagnosed in the United States in 2014–2018. Neuro Oncol 24:iii1–iii38. https://doi.org/10.1093/neuonc/noac161

6. Hoffman LM, van Zanten SEMV, Colditz N et al (2018) Clinical, radiologic, pathologic, and molecular characteristics of long-term survivors of diffuse intrinsic pontine glioma (DIPG): a collaborative report from the International and European Society for Pediatric Oncology DIPG Registries. J Clin Oncol 36: 1963. https://doi.org/10.1200/JCO.2017.75.9308
7. Schwark K, Messinger D, Cummings JR et al (2022) Receptor tyrosine kinase (RTK) targeting in pediatric high-grade glioma and diffuse midline glioma: pre-clinical models and precision medicine. Front Oncol 12:922928. https://doi.org/10.3389/fonc.2022.922928
8. Voon HPJ, Wong LH (2023) Chromatin mutations in pediatric high grade gliomas. Front Oncol 12:1104129. https://doi.org/10.3389/fonc.2022.1104129
9. Larson JD, Kasper LH, Paugh BS et al (2019) Histone H3.3 K27M accelerates spontaneous brainstem glioma and drives restricted changes in bivalent gene expression. Cancer Cell 35: 140–155.e7. https://doi.org/10.1016/j.ccell.2018.11.015
10. Patel SK, Hartley RM, Wei X et al (2020) Generation of diffuse intrinsic pontine glioma mouse models by brainstem-targeted in utero electroporation. Neuro Oncol 22:381–392. https://doi.org/10.1093/neuonc/noz197
11. du Chatinier A, Meel MH, Das AI et al (2022) Generation of immunocompetent syngeneic allograft mouse models for pediatric diffuse midline glioma. Neurooncol Adv 4:vdac079. https://doi.org/10.1093/noajnl/vdac079
12. Cordero FJ, Huang Z, Grenier C et al (2017) Histone H3.3K27M represses p16 to accelerate gliomagenesis in a murine model of DIPG. Mol Cancer Res 15:1243–1254. https://doi.org/10.1158/1541-7786.MCR-16-0389
13. Hoeman CM, Cordero FJ, Hu G et al (2019) ACVR1 R206H cooperates with H3.1K27M in promoting diffuse intrinsic pontine glioma pathogenesis. Nat Commun 10:1023. https://doi.org/10.1038/s41467-019-08823-9
14. Tomita Y, Shimazu Y, Somasundaram A et al (2022) A novel mouse model of diffuse midline glioma initiated in neonatal oligodendrocyte progenitor cells highlights cell-of-origin dependent effects of H3K27M. Glia 70: 1681–1698. https://doi.org/10.1002/glia.24189
15. Misuraca KL, Hu G, Barton KL et al (2016) A novel mouse model of diffuse intrinsic pontine glioma initiated in Pax3-expressing cells. Neoplasia 18:60. https://doi.org/10.1016/j.neo.2015.12.002
16. Barton KL, Misuraca K, Cordero F et al (2013) PD-0332991, a CDK4/6 inhibitor, significantly prolongs survival in a genetically engineered mouse model of brainstem glioma. PLoS One 8:e77639. https://doi.org/10.1371/journal.pone.0077639
17. Stewart CE, Guerra-García ME, Luo L et al (2022) The effect of Atm loss on radiosensitivity of a primary mouse model of Pten-deleted brainstem glioma. Cancers 14:4506. https://doi.org/10.3390/cancers14184506
18. Mangoli A, Wu S, Liu HQ, et al (2023) Ataxia-telangiectasia mutated (Atm) disruption sensitizes spatially-directed H3.3K27M/TP53 diffuse midline gliomas to radiation therapy. bioRxiv 2023.10.18.562892. https://doi.org/10.1101/2023.10.18.562892
19. Kanvinde PP, Malla AP, Connolly NP et al (2021) Leveraging the replication-competent avian-like sarcoma virus/tumor virus receptor-A system for modeling human gliomas. Glia 69:2059–2076. https://doi.org/10.1002/glia.23984
20. Oldrini B, Curiel-García Á, Marques C et al (2018) Somatic genome editing with the RCAS-TVA-CRISPR-Cas9 system for precision tumor modeling. Nat Commun 9:1466. https://doi.org/10.1038/s41467-018-03731-w
21. Deland K, Starr BF, Mercer JS, et al. Tumor genotype dictates radiosensitization after Atm deletion in primary brainstem glioma models. J Clin Invest 131:e142158. https://doi.org/10.1172/JCI142158
22. Weidenhammer LB, Liu HQ, Luo L et al (2023) Inducing primary brainstem gliomas in genetically engineered mice using RCAS/TVA retroviruses and Cre/loxP recombination. STAR Protoc 4:102094. https://doi.org/10.1016/j.xpro.2023.102094

Chapter 8

Preclinical Models for Assessing the Impact of Cranial Irradiation on Executive Function

Charles L. Limoli, Marie-Catherine Vozenin, and Munjal M. Acharya

Abstract

Cranial irradiation is a primary treatment modality for forestalling the growth of primary and secondary brain malignancies. While treatment of oligo metastasis has evolved toward highly precise and targeted stereotactic radiotherapy, the doses and volumes treated are still at the limit of normal brain tolerance, and the resultant impairments span a range of learning and memory and higher-order cognitive functions that adversely impact quality of life. To advance our understanding of radiation-induced cognitive dysfunction and to select interventions designed to ameliorate such deficits, behavioral testing in rodents is an indispensable tool that provides critical readouts on the functionality of the brain. Over the years, a multitude of tasks have been developed, refined, and validated that inform on the extent of the specific stressors that impact locomotion, coordination, social interaction, anxiety, depression, and learning and memory. The focus of this chapter will be on cognitive function, in part because this represents the most consequential complaint of brain tumor survivors who suffer from treatment-associated neurocognitive decline.

Key words Cranial irradiation, Neurocognitive testing, Executive function, Radiation-induced cognitive impairment

1 Introduction

Cognitive and behavioral testing, regardless of species, involves a series of carefully controlled events, where outcomes can be impacted by subtle changes in the environment, inconsistencies in testing, age, sex, and strain. While this chapter will not delve deeply into many of these issues, the methodologies presented can be applied equally across age, sex, and strain, although outcomes will be dependent in part on these factors. A variety of spontaneous exploration tasks can be implemented that test various nuances of working, episodic, recency, and associative memory functions and ultimately rely on the capability of a rodent to recognize novelty. These tests are relatively low stress, as animals ambulate about open arenas having select spatial configurations of objects and cues.

Michael Goldstein (ed.), *Translational Research Methods for Pediatric Brain Tumors*, Neuromethods, vol. 226,
https://doi.org/10.1007/978-1-0716-4754-7_8,

Examples of these tasks include Novel Object Recognition (NOR), Novel Place Recognition (NPR), Objects in Place (OIP), and Temporal Order (TO) tasks [1]. While such tasks are useful in screening and validating select stressors for their capability to elicit cognitive decline, more subtle effects often require more sophisticated testing platforms that can increase task rigor or cognitive load. Successful completion of these tasks require intact cognitive reserve and executive function skill sets that facilitate problem-solving, cognitive flexibility, and multitasking. Examples of these tasks can include a variety of touch screen paradigms such as the Psychomotor Vigilance (PVT) [2, 3] and Delayed Match to Sample (DMS) tasks [4, 5] that can test recognition reaction times, the Attentional Set-Shifting task (ATSET) that tests the capability to sequentially shift attention-based responses between one task and another [6, 7], Objects in Updated Locations (OUL) task that can test multiple associative memories [8–10], Fear Extinction memory task (FE) that requires active dissociation (unlearning) of a conditioned and unconditioned response [11] and Puzzle Box (PB) that requires select forms of problem-solving over a sequence of tasks with increasing rigor [12]. While certain tests, such as FE, involve mild levels of discomfort, other tasks can be adapted to increase task rigor (OUL, PB) without the need to invoke mildly stressful unconditioned stimuli. This chapter will focus on the methods involved in the latter three tasks.

2 Materials

1. Conditioned Stimulus: neutral stimulus.
2. Unconditioned Stimulus: mild shock to induce fear responses (e.g., foot shock).
3. Shock apparatus.
4. Behavioral chamber (contextual conditioning apparatus).
5. Video camera.
6. Puzzle box.

3 Methods

3.1 Objects in Updated Locations (OUL)

The OUL task is a cross-species task able to provide more direct comparisons between performance changes across species (mice and humans). The OUL task is an extension of the NPR paradigm, which increases task rigor from the update (NPR phase) to the test phase. This then provides additional information on multiple memory traces using a single behavioral paradigm, where test animals need to discriminate between prior overlapping associative

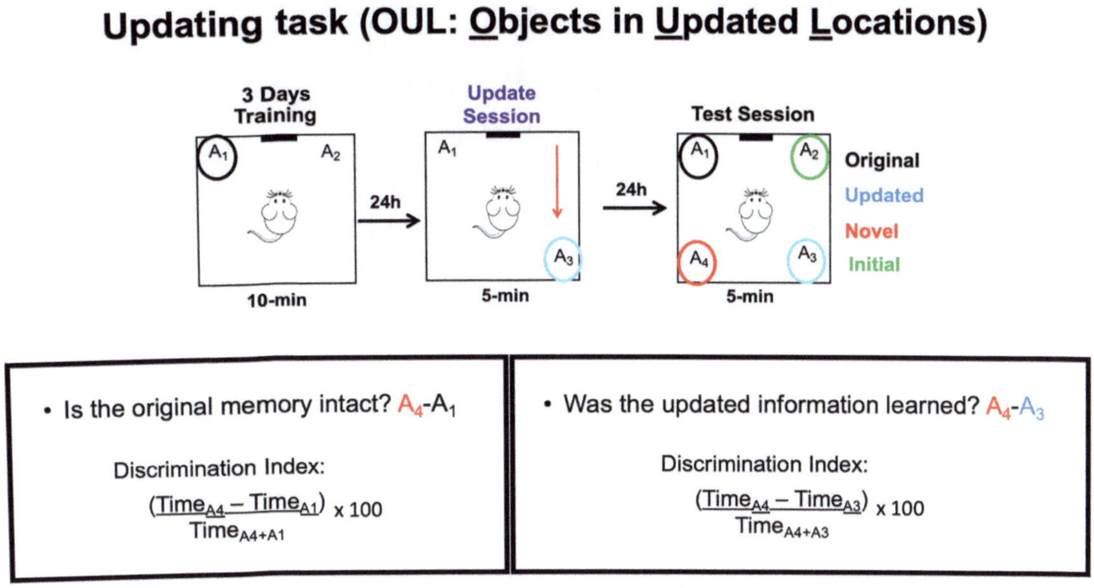

Fig. 1 Objects in Updated Locations (OUL) task

memories to recognize the most recent novel locations [8, 9]. In mice, one set of spatially positioned objects (A_1–A_4) are implemented, which represent the physical objects that they explore.

Objects are presented in pairs of locations at training (A_1, A_2, Fig. 1) and later, one of the objects can have its location moved or "updated" (A_3, Fig. 1) during further training sessions. Lastly, memory is probed by quantifying the exploration of objects presented at 4 separate locations (A_1 original, A_2 initial, A_3 updated, and A_4 novel, Fig. 1). The sequence of handling, habituating, training, and testing mice across this 2-week paradigm is shown (*see* Table 1).

As with most exploration tasks, mice are habituated to the context (5 min/day in the arena) over multiple days. Habituation time can vary depending on preference but typically transpires over 6 consecutive days. Next, mice are trained to recognize the location of two identical objects positioned at distinct locations for 10 min per day for 3–7 consecutive days (again this training period can be adjusted depending on experimental design and preference). One day following object location memory training, mice are given a 5-min update session in which one object is moved to a new location (A_3, day 12), and the amount of time spent exploring the relocated vs. familiar object location is then evaluated. The next day, mice are tested for 5 min with all four identical objects (A_{1-4}) in distinct locations to determine whether the original memory was updated to incorporate the new location information. Time spent exploring the novel object location compared with the updated, original, and moved object location can then be examined. In this

Table 1
The sequence of handling, habituating, training, and testing mice across the 2-week paradigm

Day 1	Day 2	Day 3	Day 4	Day 5	Day 6	Day 7	Day 8	Day 9	Day 10	Day 11	Day 12	Day 13
Handle	Handle	Handle	Handle									
Acclimatization		Habituate (no objects)	Habituate (no objects)	Habituate (no objects)	Habituate (no objects)	Habituate (no objects)	Habituate (no objects)	Train	Train	Train	Update (1st olm)	Test (4 objects)
		5 min	5 min	5 min	5 min	5 min	5 min	10 min	10 min	10 min	5 min	5 min
								Leave in ante room 30–60 min before start	Leave in ante room 30–60 min before start	Leave in ante room 30–60 min before start	Leave in ante room 30–60 min before start	Leave in ante room 30–60 min before start

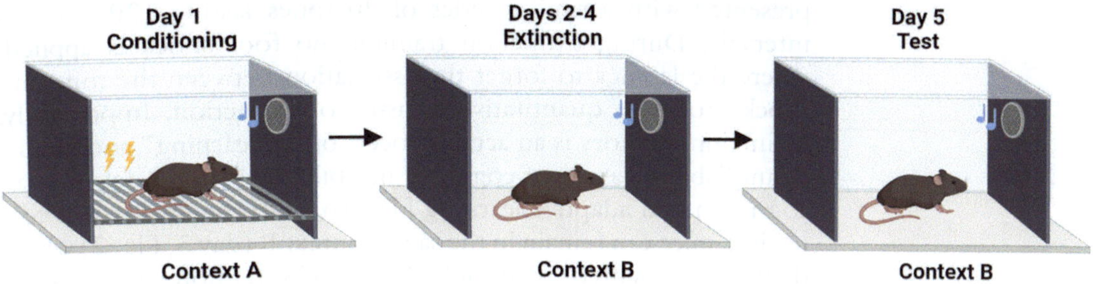

Fig. 2 Fear extinction memory task

instance, comparing the novel vs updated (A_4–A_3), the novel vs initial (A_4–A_2), or the novel vs original (A_4–A_1) can be quantified using the discrimination index (*see* Fig. 1) and can be used to determine the extent that the test animal retains a memory for the updated location, the initial configuration or the original training information respectively. Since animals with relatively normal brain function will exhibit a tendency to explore novelty, the discrimination indices of both the original and updated memory are typically diminished in animals with impaired cognition resulting from radiation exposure.

Fear Extinction

The fear extinction memory task is an expanded task following the initial phase of the fear conditioning response (freezing) that involves relearning [13, 14]. In both paradigms, fear acquisition is obtained during a conditioning phase in which a series of 3 tone-shock pairings spaced 120 s apart are delivered on day 1 (Context A, Fig. 2) to ensure learning. For the auditory conditioned stimulus (CS), each 80 dB, 16 kHz tone lasts 120 s, followed by the foot shock unconditioned stimulus (US, 0.6 mA, 1 s) that results in a "freezing behavior" as the animal anticipates a shock when the tone is played. The number of tone-shock pairings can be increased if the average amount of time freezing does not reach ~50% of the time the animal is monitored after the final tone-shock pairing (during the final 120 s interval). In fear conditioning, the animals are then exposed to a similar series of tones the day after in the absence of the foot shock and monitored for freezing behavior which indicates an intact fear response. Animals that are irradiated or subjected to other stressors typically exhibit reduced freezing that provide a quantitative measure of cognitive decline not reliant on exploratory activity.

In fear extinction, the test is adapted to provide a measure of cognitive flexibility, where during subsequent extinction training (days 2–4, Fig. 2) the animals are placed in a different context (Context B), where after a brief 2 min habituation, animals are presented with a similar series of 20 tones lasting 120 s at 5 s intervals. During extinction training, no foot shock is applied, where the latency to forget the association between the tone and shock provide a quantitative measure of extinction. Importantly, extinction memory is an active process of "unlearning" and gives a quantifiable measure of cognitive flexibility, which models human cognition and adaptive learning [15, 16]. On the final day of fear testing, mice can remain in the same context B (day 5, Fig. 2) where they are presented with only 3 non-US reinforced CS tones (16 kKz, 80 dB, lasting 120 s) at a 2-min interval or switched to their original context A. Typical changes in context involve both changes in scent (vinegar vs almond) and the floor plate (electrical grids vs no grids, and acrylic floor). When the testing phase is conducted in the same context B (as shown in Fig. 2), this particular paradigm interrogates both learning and memory elements dependent on intact hippocampal and medial prefrontal cortex (mPFC) function. If, however, the entire testing paradigm was conducted in the new context (new odor and spatial cues), then the test would be reliant on the mPFC. It is also worth noting that the testing phase can be conducted virtually any time after extinction training (days to months), where persistent freezing or the inability to unlearn the association between the CS and US will depend on many factors, including the nature of the stressor that disrupts the fear circuitry in the brain.

The fear response of animals can be recorded and analyzed by a system involving a FreezeFrame module controlled by computers. The FreezeFrame module is a video-based system for fear conditioning/extinction that has now become the standard in its field. With such systems, one can detect the minute movements of grooming, sniffing, turning, and rearing, and importantly, such systems allow reliable assessments of extinction memory under a reasonable time frame.

3.2 Puzzle Box Test

Cranial irradiation and cancer-related cognitive impairments (CRCI) are often associated with higher-order cognitive flexibility and executive function-related deficits in survivors [17]. Puzzle box (PB) protocol tests executive function by assessing animal's learning, problem-solving, and memory function abilities, which are reliant on intact medial prefrontal cortex (mPFC) and partially hippocampal function [12, 18, 19]. PB task can also be extended to measure short- and long-term memory functions. Briefly, mice are placed in a plexiglass compartmentalized light and dark box arena. The brightly lit (250–350 lux), larger white compartment (40 × 25 × 30 cm; l × w × h), and a black lid-enclosed, dark compartment (<7 lux, 20 × 25 × 30 cm) are separated by a black plexiglass divider wall (25 × 30 cm) with an underpass opening in the bottom of the arena (43 × 43 mm) to allow the passage of mice. This underpass can also be constructed as a trench (optional). A red plexiglass mouse igloo or tube is kept in the dark compartment. The placement of mice in the brightly lit white compartment motivates their preference to quickly migrate into the dark compartment (goal zone) due to their nocturnal nature. Mice are first acclimatized in the light-dark box for the first day. On the subsequent days, mice are challenged with a number of obstacles (including a tray, bedding or a paper plug) with increasing rigor of difficulty. Mice have to solve how to pass through these barriers to reach the dark goal zone. Sequentially introducing obstacles of increasing difficulty and rigor tests for learning, short- and long-term memory, and problem-solving abilities. The puzzle box trials, obstacle training, and testing are conducted, as shown in Table 2.

Mice are first acclimatized (T0) in the light-dark box compartment without any obstacles on Day 1 for 15 min. This phase also allows for the determination of anxiety-like behavior [20] by assessing the latency of reaching the dark compartment and the number of transitions between the light and dark compartments. Cranial irradiation has been shown to increase anxiety-like behavior on the light-dark box test [21]. On Day 2, during the T1 trial, mice explore the arena without obstacles for 10 min. During T2 and T3 trials, a channel tray is placed within the doorway.

Table 2
Results of the puzzle box trials, obstacle training, and testing

Day	Trial	Task	Obstruction	Time (s)
1	T0	Acclimatization	Open door with no obstructions	900
2	T1	Learning	Open door with no obstructions	600
	T2	Learning	Open channel tray within the doorway	300
	T3	Learning with an obstacle	Open channel tray within the doorway	300
3	T4	Memory and problem-solving ability	Open channel tray within the doorway	300
	T5	Problem-solving ability	Channel tray filled with corncob bedding	300
	T6	Learning with an obstacle	Channel tray filled with corncob bedding	300
4	T7	Memory and problem-solving ability	Channel tray filled with corncob bedding	300
	T8	Problem-solving ability	Paper tissue plug in doorway	300
	T9	Problem-solving ability	Paper tissue plug in doorway	300
2–3 weeks later (optional)	T10	Long-term memory recall	Paper tissue plug in doorway	300

Mice learn to pass through the doorway to reach the goal zone. On Day 3, for trial T4, the channel tray trial is reintroduced (similar to T3), which tests the intact long-term memory function of the previously learned task. Subsequently, for T5 and T6 trials (burrowing puzzle), the doorway with the channel tray is filled with corncob bedding, and mice have to dig through this obstacle to reach the dark compartment. This trial tests learning and problem-solving skills. On Day 4, the T7 trial (similar to T6) tests the previous day's obstacle-solving memory in mining through the bedding-blocked doorway. Subsequently, trials T8 and T9 (plug puzzle) increase the obstacle difficulty by blocking the doorway (no channel tray) using a paper plug that mice have to pull or push through to enter the dark goal zone. To determine long-term memory recall, an optional trial, T10, can be conducted 2 to 3 weeks after the completion of T9. In this trial, mice will be introduced to the same arena with a paper tissue plug blocking the doorway. All trials are recorded by the ceiling-mounted camera. Either automated or manual scoring procedures can be used to assess the latency (seconds) to reach the target zone and the inability to solve the puzzle. A simplified schematic of this test is shown in Fig. 3.

In comparison with other executive function tasks, including ATSET and five-choice serial reaction time (5CSRT) task, the PB task requires relatively less duration, number of trials and sessions per day, the time needed to train animals, and simpler arena or apparatus requirements [12]. PB task has shown reliability in wild-type and neurodegenerative disease mouse models [12, 22]. However, animal models with higher fatigue levels, neophobic behavior, and reduced motor function may need additional validation for executive dysfunction. Notably, a number of CRCI models have been characterized using the PB task, showing the detrimental impact of cancer chemotherapies on executive function [23–25]. The PB task is relatively noninvasive and low stress. It does not require extensive training, food or water restrictions, or aversive experience (e.g., foot shocks) to determine the executive function that relies on the intact mPFC- and hippocampal function-dependent goal-directed objective (reaching the dark zone) and motor function.

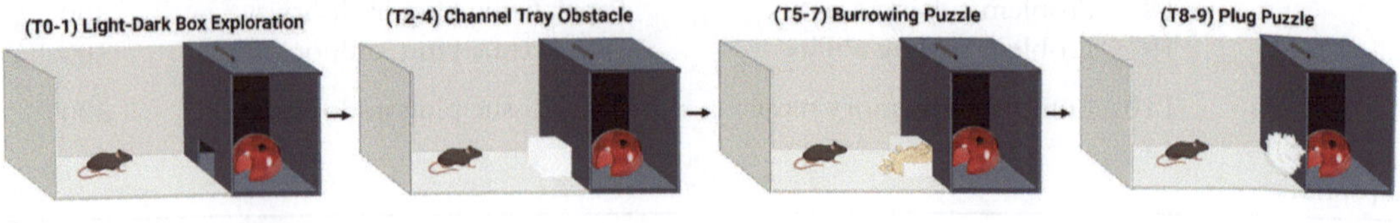

Fig. 3 Puzzle Box test

4 Conclusions

Cognitive testing in rodents provides foundational insights into the locoregional and neural basis of cognition and is useful for quantitative assessment of behavioral performance. While many testing options are available, for oncology-related studies, the selection of the testing paradigm must be based upon the hypothesis underlying the study and provide consistent results in tumor-free and tumor-bearing animals. In the context of radiation-induced cognitive decrements, for which the timing and nature of the induction factor are clearly identified, significant interindividual variations within cohorts can be observed even in tumor-free animals. This requires adaptation of the sample size using power calculation to obtain significant results. The selection of the relevant timing to conduct the behavioral studies is also essential and must be adapted according to the strain of animal, the volume of brain exposed, and the dose of irradiation, keeping in mind that neurocognitive decrements are delayed disorders, typically transpiring several months postirradiation. Selection of the appropriate rodent model (mouse vs. rat, strain) is also an important consideration, as differences between the strains of each species can impact task selection and behavioral outcomes. While rats are smarter and can be tested for more complex problem solving, genetically engineered mice afford enhanced mechanistic insight. The impact of sex is another possible confounding factor that needs to be mitigated using testing strategies providing consistent results in males and females and/or specifically investigated to understand how sex differences, influenced by hormones, genetics, and environmental factors, can affect cognitive abilities. In summary, the panel of tests described in this chapter provide a selection of reliable and translational behavioral assays that offer the possibility to interpret the results in a broader clinical context taking ethical concerns in considerations. While rodent behavioral tests are intrinsically less complex than those conducted in humans, they are useful to explore mechanisms and/or mitigating strategies that can later be investigated in human studies.

Acknowledgments

This work was supported by the National Institutes of Health (NIH) awards P01 CA244091 (CLL, MCV), R01 CA254892 (CLL, MCV), and R01 CA251110 (MMA).

Conflict of Interest The authors have no conflicts of interest to declare that are relevant to the content of this chapter.

References

1. Drayson O, Vozenin M-C, Limoli C (2023) A rigorous behavioral platform for the assessment of radiation-induced neurological outcomes. Method Molec Biol, vol 180, pp 177–197
2. Basner M et al (2018) Repeated administration effects on psychomotor vigilance test performance. Sleep 41(1). https://doi.org/10.1093/sleep/zsx187
3. Desai RI et al (2023) Complex 33-beam simulated galactic cosmic radiation exposure impacts cognitive function and prefrontal cortex neurotransmitter networks in male mice. Nat Commun 14(1):7779
4. Winocur G, Hasher L (2004) Age and time-of-day effects on learning and memory in a non--matching-to-sample test. Neurobiol Aging 25(8):1107–1115
5. Robbins ME et al (2011) A model for assessing cognitive impairment after fractionated whole-brain irradiation in nonhuman primates. Radiat Res 175(4):519–525
6. Jewell JS et al (2018) Exposure to </=15 cGy of 600 MeV/n (56)Fe particles impairs rule acquisition but not long-term memory in the attentional set-shifting assay. Radiat Res 190(6):565–575
7. Britten RA et al (2022) Dissecting differential complex behavioral responses to simulated space radiation exposures. Radiat Res 197(3):289–297
8. Kwapis JL et al (2020) Aging mice show impaired memory updating in the novel OUL updating paradigm. Neuropsychopharmacology 45(2):337–346
9. Wright DS, Bodinayake KK, Kwapis JL (2020) Investigating memory updating in mice using the objects in updated locations task. Curr Protoc Neurosci 91(1):e87
10. Alaghband Y et al (2023) Uncovering the protective neurologic mechanisms of hypofractionated FLASH radiotherapy. Cancer Res Commun 3(4):1–13
11. VanElzakker MB et al (2014) From Pavlov to PTSD: the extinction of conditioned fear in rodents, humans, and anxiety disorders. Neurobiol Learn Mem 113:3–18
12. Ben Abdallah NM et al (2011) The puzzle box as a simple and efficient behavioral test for exploring impairments of general cognition and executive functions in mouse models of schizophrenia. Exp Neurol 227(1):42–52
13. Parihar VK et al (2016) Cosmic radiation exposure and persistent cognitive dysfunction. Sci Rep 6:34774
14. Parihar VK et al (2018) Persistent nature of alterations in cognition and neuronal circuit excitability after exposure to simulated cosmic radiation in mice. Exp Neurol 305:44–55
15. Milad MR, Quirk GJ (2002) Neurons in medial prefrontal cortex signal memory for fear extinction. Nature 420(6911):70–74
16. Sotres-Bayon F et al (2012) Gating of fear in prelimbic cortex by hippocampal and amygdala inputs. Neuron 76(4):804–812
17. Makale MT et al (2017) Mechanisms of radiotherapy-associated cognitive disability in patients with brain tumours. Nat Rev Neurol 13(1):52–64
18. Ruddy RM, Adams KV, Morshead CM (2019) Age- and sex-dependent effects of metformin on neural precursor cells and cognitive recovery in a model of neonatal stroke. Sci Adv 5(9):eaax1912
19. Milenkovic M, Mielnik CA, Ramsey AJ (2014) NMDA receptor-deficient mice display sexual dimorphism in the onset and severity of behavioural abnormalities. Genes Brain Behav 13(8):850–862
20. Bourin M, Hascoet M (2003) The mouse light/dark box test. Eur J Pharmacol 463(1–3):55–65
21. Montay-Gruel P et al (2019) Long-term neurocognitive benefits of FLASH radiotherapy driven by reduced reactive oxygen species. Proc Natl Acad Sci USA 116(22):10943–10951
22. Galsworthy MJ et al (2005) Assessing reliability, heritability and general cognitive ability in a battery of cognitive tasks for laboratory mice. Behav Genet 35(5):675–692
23. Kim T et al (2024) Gamma entrainment using audiovisual stimuli alleviates chemobrain pathology and cognitive impairment induced by chemotherapy in mice. Sci Transl Med 16(737):eadf4601
24. Chiu GS et al (2018) Nasal administration of mesenchymal stem cells restores cisplatin-induced cognitive impairment and brain damage in mice. Oncotarget 9(85):35581–35597
25. Chiang ACA et al (2020) Bexarotene normalizes chemotherapy-induced myelin decompaction and reverses cognitive and sensorimotor deficits in mice. Acta Neuropathol Commun 8(1):193

Chapter 9

A Rapid, Reliable, and Reproducible Cisterna Magna Injection Approach for the Treatment of Leptomeningeal Metastatic Brain Tumors

Zhenhua Xu, Najiba Murad, and Yanxin Pei

Abstract

Leptomeningeal metastases (LM) in the setting of pediatric brain tumors are associated with particularly poor prognosis. Current therapy for pediatric brain tumors includes surgery, radiation, and chemotherapy. However, LM are not amenable to surgical resection and frequently do not respond to radiation. Moreover, most chemotherapeutic drugs are unable to cross the blood-brain barrier or blood-cerebrospinal fluid (CSF) barrier to reach LM. Here, we describe a mouse model of pediatric brain tumors that develop LM and an easy, reproducible approach to deliver therapeutic agents to treat LM. We intrathecally injected a CSF tracer via the cisterna magna in mice. After a single injection, CSF tracer influx was observed along the leptomeninges throughout the entire central nervous system and persisted for at least 6 h. Daily injection did not result in apparent toxicity or impact animal weight. The results of this study highlight the utility of a mouse model of metastatic pediatric brain tumors as well as the potential application of a repeated intrathecal injection technique to deliver therapeutic agents to treat metastatic brain tumors.

Key words Brain tumor, Leptomeningeal metastases, Medulloblastoma, CSF, Cisternal magna

1 Introduction

Leptomeningeal metastases (LM) are common in patients with advanced brain cancer [1–3], breast cancer [4, 5], lung cancer [6], melanoma [7, 8], and leukemia [9]. The presence of LM uniformly confers a poor prognosis and negatively impacts patient quality of life. Because meaningful surgical resection is impossible due to the diffuse nature of LM, the standard of care for leptomeningeal metastatic disease involves radiation and chemotherapy [10]. However, extensive irradiation of the entire central nervous system often causes severe treatment-related side effects, particularly for young patients. Moreover, many chemotherapeutic drugs cannot readily cross the blood-brain barrier and blood-cerebrospinal fluid (CSF) barrier, limiting their distribution to

Michael Goldstein (ed.), *Translational Research Methods for Pediatric Brain Tumors*, Neuromethods, vol. 226,
https://doi.org/10.1007/978-1-0716-4754-7_9,

LM upon intraperitoneal, intravenous, or oral administration [11]. For drugs that are capable of permeating these barriers, achieving effective pharmacological concentrations through simple diffusion from the blood to the LM remains a challenge [12, 13]. *Intrathecal* injection, which allows direct drug *delivery* to the CSF, can circumvent these barriers and is widely used for the treatment of LM [14]. Intrathecal injection can be performed through either lumbar puncture or implantation of catheters/cannulas into the ventricles or cisterna magna [15–18]. However, the surgery that is required to perform these procedures in preclinical models is inconvenient and may cause undue stress to animals [16, 18–22]. These procedures also carry the risk of post-operative injection injury [16, 23]. Thus, the goal of this study was to develop a noninvasive approach to repeatedly and accurately deliver agents directly to the CSF to facilitate the treatment of LM in preclinical mouse models.

Medulloblastoma, one of the most common malignant pediatric brain tumors, is a significant cause of childhood morbidity and mortality [24, 25]. Approximately one-third of patients with malignant medulloblastoma present with metastases, and nearly all have metastatic disease at relapse [26, 27]. Medulloblastoma metastases are found almost exclusively along the leptomeninges on the surface of the brain and spinal cord [5, 28, 29]. Therefore, intrathecal drug delivery shows promise as an effective technique for the treatment of metastatic medulloblastoma. Here, we describe a protocol to generate a metastatic medulloblastoma mouse model as well as a simple and reliable intrathecal injection approach to deliver agents to the CSF in mice. Because surgery is not required, this approach is less invasive than other methods and allows animals to recover more rapidly.

2 Materials

1. 33-gauge Neuros syringe, 10 μL.
2. 4% paraformaldehyde solution in PBS.
3. 5 μL Microliter Syringe Model 7105 KH, Knurled Hub Needle, 24 gauge, 2.75 in, point style 3.
4. aCSF.
5. Bard-Parker Sterile Disposable Scalpels #10.
6. Buprenorphine HCL injectable, 0.3MG/ML, C3.
7. Butler Schein Puralube Nonantibiotic ophthalmic ointment.
8. Cryostat.
9. DAPI, FluoroPure™ grade.
10. Deltaphase Isothermal Pad.

11. Digital 3-Axes Manipulator Arm, left.
12. Digital Conversion for Stoelting Manipulator Arm, left.
13. Digital Just for Mouse Stereotaxic.
14. DPBS, no calcium, no magnesium.
15. Dumont #5 Forceps.
16. Dumont #7 Forceps.
17. Ethanol.
18. Extra Narrow Scissors.
19. Fisherbrand Sterile Alcohol Prep Pad.
20. Fluorescein isothiocyanate–dextran.
21. Fluorescence microscope.
22. Fluorescence stereo microscope.
23. Gas Anesthesia Mask for Stereotaxis.
24. Germinator 500.
25. Graefe Forceps, 10.2 cm.
26. Heat Lamp with table clamp.
27. Induction chamber, anesthesia accessories, small animal, 5.25 L.
28. Isoflurane solution.
29. IVIS Lumina III In Vivo Imaging System.
30. Liquivet (Rapid) Tissue Adhesive.
31. Neurocult stem cell medium and proliferation supplement.
32. Quintessential Stereotaxic Injector (QSI).
33. Restrainer & Split Shield.
34. RS 2000 Biological System Irradiator.
35. Sterile Cotton-Tipped Applicators.
36. Sucrose.
37. Surgical Clipper, Cordless, Rechargeable w/#40 blade.
38. Tissue-Plus OCT Compound.
39. XenoLight D-Luciferin Potassium Salt.

3 Methods

All animal procedures were performed according to NIH guidelines and were approved by the Research Animal Facility Committee, Children's Research Institute (Protocol #30351).

3.1 Generation of a Mouse Model of Metastatic Medulloblastoma

Note: Because this is a recovery surgery, procedures should be performed under sterile conditions. All surgical tools should be sterilized by autoclaving or with a hot bead sterilizer.

3.1.1 Stereotactic Implantation of Medulloblastoma Tumor Cells

1. Transfect tumor cells with mCherry and luciferase reporters. To transfect cells, culture 2 × 10^5 patient-driven xenograft medulloblastoma cells in a 24-well plate with 360 μL neural stem cell medium and proliferation supplement. Add 40 μL of lentivirus encoding both mCherry and luciferase in one vector to the culture medium without polybrene, and culture overnight.
2. After overnight culture, collect cells into a microcentrifuge tube by pipetting.
3. Centrifuge cells for 5 min at 220×g and room temperature. Remove supernatant carefully with a micropipette and discard supernatant into a 5% bleach solution.
4. Wash once by resuspending the pelleted cells in 500 μL fresh neural stem cell medium and centrifuging for 5 min at 220×g and room temperature.
5. Remove the supernatant and resuspend cells in 16 μL neural stem cell medium. Keep the tube on ice for the duration of the transplantation procedure.
6. Anesthetize an immunodeficient mouse (4–6 weeks old) by placing the animal in an isoflurane induction chamber (3% isoflurane, 97% air, 250 CC/min) until the toe pinch reflex ceases.
7. Load the mouse in the prone position on a stereotactic frame over a heating pad connected to an anesthesia gas mask to keep the mouse under 1.5–2% isoflurane anesthesia.
8. Disinfect the head of the mouse with 70% ethanol and apply nonantibiotic ophthalmic ointment to keep the eyes moist while the mouse is anesthetized.
9. Using a sterile #10 scalpel, make an approximately 12 mm incision in the skin at the midline of the scalp over the position of the cerebellum. Expose the skull by pushing aside the muscle tissue over the lambdoid suture with sterile fine forceps.
10. Determine the coordinates for injection into the cerebellum using a mouse brain atlas (1 mm lateral to the midline). Using the beveled end (sharp point) of a sterile 18-gauge needle to puncture the skull, create a small hole with a diameter of approximately 0.5 mm in the skull. A dissection microscope should be used to visualize the injection site.

11. Load cells (5×10^4 in 4 μL neural stem cell media) into a 5 μL microliter syringe with an unbeveled 24-gauge needle, mount the syringe on a micromanipulator, and introduce the needle through the hole in the skull into the brain at a depth of 2.2 mm.
12. Inject cells using an electronic pump at a flow rate of 3 μL/min. After injection, leave the needle in place for 1 min to avoid backflow of the cells or fluid.
13. Withdraw the needle slowly and remove the mouse from the frame.
14. Close the skin with tissue adhesive or sutures.
15. Place the mouse in a new cage containing bedding. Place the cage under a heat lamp to keep the animal warm. Repeat **steps f–n** for each mouse.
16. Monitor the animals until they have completely recovered from anesthesia and are fully mobile (approximately 2–3 min). Continue to monitor the animals for an additional 30 min postoperatively.
17. Administer 0.05–0.1 mg/Kg buprenorphine subcutaneously as postoperative analgesia. Continue to administer buprenorphine every 12 h for 48 h after the procedure to manage any pain from the injection. Note: guidelines for postoperative pain management vary between institutions. Please consult with your institution to ensure proper pain management.
18. Monitor tumor growth once a week using a bioluminescent imaging system.

3.1.2 Irradiation of Tumor-Bearing Mice

1. Irradiate the tumor-bearing animals when the luciferase signals reach up to $1 \times 10^{6-7}$ radiance per second. Note: The length of time for animals to develop tumors depends on the number of tumor cells, but typically, animals develop tumors within 3–4 weeks.
2. Prewarm the irradiator following the manufacturer's instructions.
3. Restrain the mouse in a lead shielding without anesthesia. The shielding is composed of two lead boxes with a narrow open space between, allowing radiation exposure to only the brain and spine while shielding the rest of the body.
4. Place the restrained mouse into the irradiator.
5. Irradiate the mouse with 2 Gy per dose at a rate of approximately 120 rads/min.
6. After irradiation, return the mouse to its original cage.
7. Repeat irradiation every other day for a total cumulative dose of up to 18 Gy.

8. Image mice once per week with a bioluminescent imaging system to monitor the development of metastases. Metastases typically develop within 3–4 weeks following the first dose of radiation.

3.1.3 Identifying Metastases

1. Perfuse mice intracardially with 0.01 M phosphate buffer saline (PBS) followed by 4% paraformaldehyde (PFA) in 0.01 M PBS at a rate of 5 mL/min.
2. Use fine forceps to remove the brain and spinal cord and place in cold PBS.
3. Use a fluorescence microscope to examine the location of mCherry$^+$ tumor cells and acquire images.
4. Fix the tissues in 4% PFA at 4 °C overnight, then dehydrate the tissues with 25% sucrose overnight and embed in tissue embedding medium. Freeze blocks in a dry ice bath with 100% ethanol and store embedded tissue at −80 °C.
5. Use a cryostat to cut 12 μm sections of brain and spinal cord.
6. Stain the sections with 4′,6-diamidino-2-phenylindole (DAPI). Examine the metastatic tumor cells in each section under a fluorescence microscope and acquire images.

3.2 Intrathecal Injection

Note: Because this is a recovery surgery, procedures should be performed under sterile conditions.

3.2.1 Intrathecal Injection of CSF Tracer

1. Prepare a 10 μL Neuros syringe by attaching a 33-gauge beveled needle and stopper. Set the stopper to 3 mm and wash the needle 10 times with sterile deionized water, then 10 times with sterile PBS.
2. Anesthetize a mouse (4–6 weeks old) by placing the animal in an isoflurane induction chamber (2.5–3% isoflurane, 97% air, 250 CC/min) until the toe pinch reflex ceases.
3. Load the mouse in the prone position on a stereotactic frame over a heating pad connected to an anesthesia gas mask to keep the mouse under 1.5–2% isoflurane anesthesia.
4. Restrain the head of the mouse between the two ear bars. Use a tooth bar to align and support the front of the head.
5. Adjust the height of the ear bars and the tooth bar to make sure the line connecting the most prominent aspects of the cranium and the spine forms an angle of 15° with the horizontal line. In this position, the cisternal magna is nearly the highest point of the mouse's body.
6. Shave the head and neck of the mouse, remove fur, disinfect the head with alcohol scrubs, and apply sterile ophthalmic ointment to keep the eyes moist during anesthesia.

7. Prepare CSF tracer by dissolving fluorescein isothiocyanate (FITC)-dextran-500 in artificial CSF (aCSF) at a concentration of 0.25% w/v. Load 3 μL CSF tracer into the 10 μL Neuros syringe.
8. Place the syringe onto the syringe holder in the micromanipulator connected to the stereotactic frame at an angle of 35°.
9. Palpate the space between the occiput and C1 vertebrae with either an index finger or a cotton-tipped swab. These two anatomical landmarks are used to determine the midline of the cisterna magna. An indentation is observed, which is used to define the puncture site. Mark the puncture site with a marker.
10. Insert the needle into the puncture site to a depth of 2.8 mm at a 35° angle, and inject 3 μL CSF tracer at a rate of 3 μL/min using an electronic infusion pump.
11. Gently remove the needle and allow the mouse to recover in a fresh cage with new bedding under a heat lamp.
12. At 30 min or 6 h after injection, perfuse mice intracardially with 0.01 M PBS, followed by 4% PFA in 0.01 M PBS at a rate of 5 mL/min.
13. Remove the entire brain and spinal cord and check for any evidence of brain tissue injury. Acquire images of the brain and spinal cord under a fluorescence microscope.
14. Postfix the brain and spinal cord in 4% PFA overnight, and cryoprotect with 25% sucrose in 0.01 M PBS at 4 °C for 24 h.
15. Section the brain and spinal cord tissue into 12 μm-thick coronal or sagittal sections using a cryostat. Stain the slides with DAPI, and examine the intracranial distribution of the CSF tracer.
16. To determine whether mice can tolerate multiple intrathecal injections, intrathecally inject aCSF for 3 consecutive days, followed by a rest period of 4 days, then inject for another 3 consecutive days.
17. Monitor the mice and record their weight daily.

3.3 Representative Results

We first evaluated metastases development after irradiation in a patient-derived xenograft mouse model. Without irradiation, animals developed large tumors in the cerebellum (Fig. 1a–e). In contrast, irradiation ablated the majority of the tumor cells in the cerebellum but facilitated the development of metastases along the leptomeninges (Fig. 1f–j).

We then characterized the movement of CSF fluorescent tracer along the leptomeninges after infusion via the cisterna magna. Tracer movement within the central nervous system was evaluated by fluorescence microscopy. We observed that FITC-dextran-500

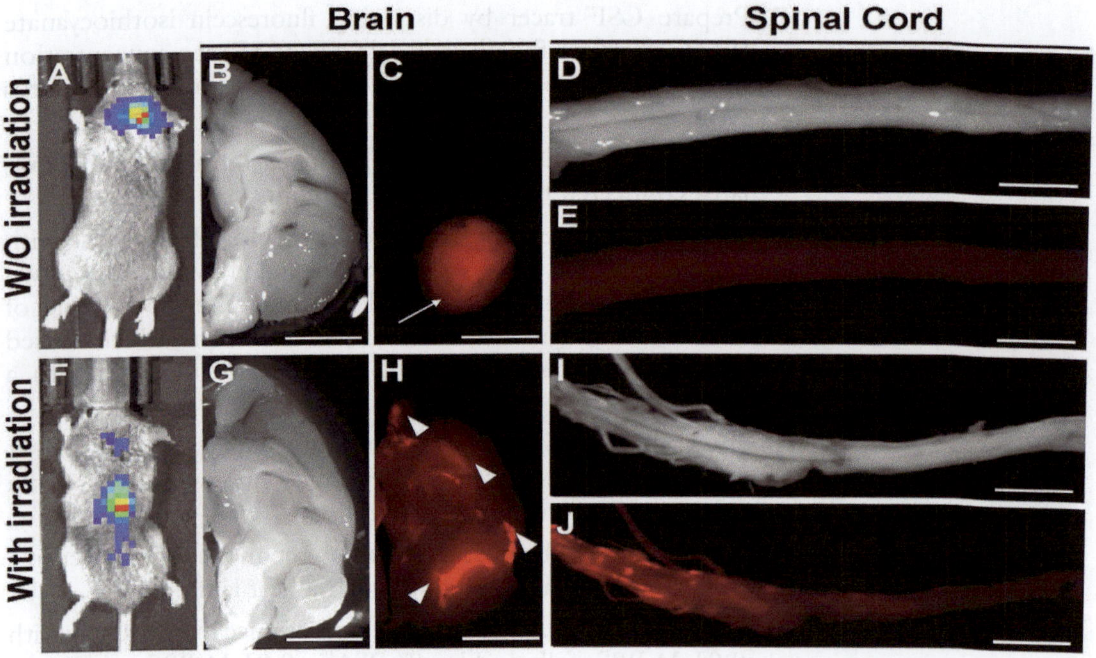

Fig. 1 Generation of metastatic medulloblastoma mouse model. Representative images of the whole mouse (**a**, **b**), brain (**b**, **c**, **g** and **h**) or spinal cord (**d**, **e**, **i**, and **j**) from the tumor-bearing mice treated without (**a**–**e**) or with craniospinal irradiation (**f**–**j**). Without irradiation, animals develop large tumors in the cerebellum (arrow) at 4–5 weeks posttransplantation but rarely develop metastases. In contrast, irradiation leads to an increase in metastases along the leptomeninges (arrowheads) at 3–4 weeks after the first dose of irradiation. mCherry indicates tumor cells. Scale bars, 3 mm

rapidly moved along the pial surface and perivascular pathways (Fig. 2). Under the same imaging conditions, no fluorescence was observed in uninfused brain tissue (data not shown). Analysis of FITC-dextran-500 fluorescence in the whole brain and spinal cord revealed that CSF tracer influx was strong at 30 min post-infusion (Fig. 2a–d), but declined at 6 h (Fig. 2e–h) ($n = 3$ animals in each group). This observation is consistent with previous studies showing that the influx of CSF tracer peaks approximately 30 min post-infusion and dissipates at later time points [16, 17]. To quantify tracer movement into different areas of the brain, we evaluated four regions of interest, including the olfactory bulb, ventricle, cerebellum, and spinal cord. Figure 2i–n shows representative images of brain or spinal cord at 30 min after infusion. CSF tracer was observed in each of these regions, suggesting CSF tracer movement via CSF circulation. CSF tracer circulation was observed in all 10 mice that received a single cisterna magna injection. No tissue damage surrounding the puncture site was detected in these animals, suggesting that this technique is safe and reliable. We further evaluated the effect of chronic intracisternal injections on animal weight. We performed intracisternal aCSF infusion in 6 animals on

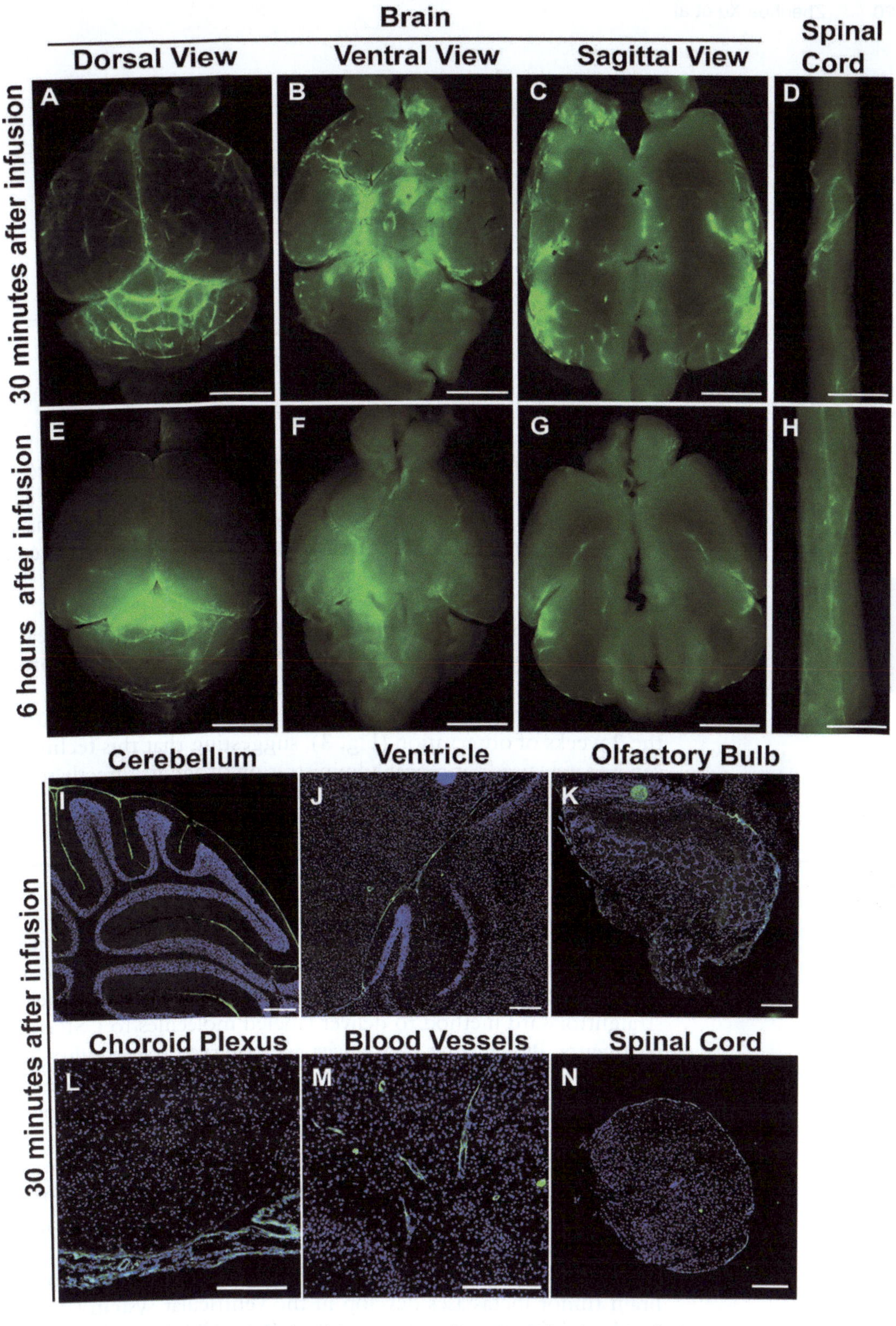

Fig. 2 Injection of CSF tracer into the cisterna magna. Representative images of whole mouse brains or spinal cords at 30 min or 6 h after cisterna magna injection with the fluorescent tracer FITC-dextran-500. Brains were imaged from dorsal (**a**, **e**), ventral (**b**, **f**) and sagittal (**c**, **g**) aspects. Spinal cord images (**d**, **h**). (**i**–**n**) Representative sagittal sections of the brain and transverse spinal cord sections were counterstained with DAPI at 30 min after cisterna magna injection. Green indicates FITC-dextran-500. Blue, DAPI. Scale bars, **a**–**h**, 3 mm, **i**–**n**, 200 μm

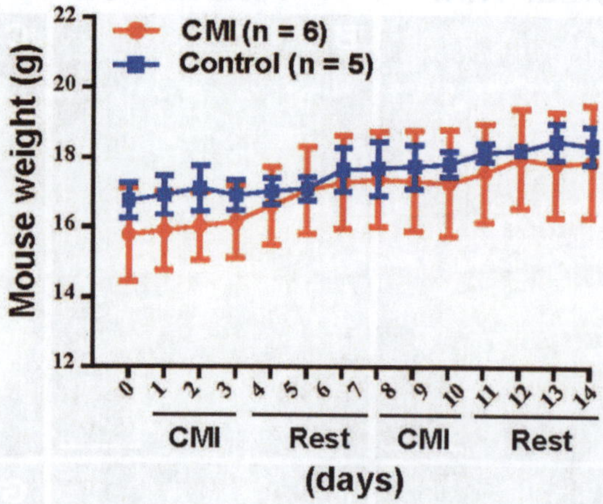

Fig. 3 Change in animal weight upon multiple injections of aCSF into the cisternal magna. aCSF was injected intrathecally for 3 consecutive days, followed by a rest period of 4 days, then another 3 consecutive days. Animals that did not receive injection served as the controls. Animal weight was recorded daily for 2 weeks. CMI, cisterna magna injection

3 consecutive days followed by a rest period of 4 days, then repeated this process for an additional week. Animal weight was recorded daily for 2 weeks.

Cisterna magna injection did not affect animal weight during the 2 weeks of observation (Fig. 3), suggesting that this technique can be safely used for repeat administration of agents directly to the intrathecal space.

4 Notes

In this report, we present a detailed protocol to generate a metastatic brain tumor mouse model and perform intrathecal injection into the cisterna magna. This injection approach provides a straightforward method to deliver labeled molecules to CSF compartments. By injecting CSF tracer, we observed tracer influx into different regions of the brain and spinal cord (see Fig. 9.2). Agents with differing molecular sizes may have distinct diffusion properties; molecules that are smaller in size often have a faster diffusion rate than molecules that are larger in size [17, 30, 31]. In our experiment, we selected FITC-dextran 500 CSF tracer. With a molecular weight of 500 kDa, the size of this tracer is much larger than that of most small-molecule drugs, which usually have molecular weights of less than 500 Da [32, 33]. Using our approach, brain tumor metastases develop in the ventricular system (lateral, third, and fourth ventricles) (see Fig. 9.1). CSF is produced predominantly by the choroid plexus in the lateral, third, and fourth ventricles and is distributed over the surface of the brain [13]. Therefore, intrathecal injection can allow agents to flow through the entire ventricular system where metastatic brain tumor cells are located. Our approach demonstrates the feasibility of a simple technique with potential clinical application to deliver therapeutic agents against brain tumor LM.

Our study represents substantial modifications of previous methods [34, 35] to allow accurate and replicable drug delivery into the cisterna magna in mice. Our approach uses direct visualization of the puncture site through removal of the hair as opposed to transcutaneous blind skin incision and injection. Our technique also improves upon previous methods by placing the injection needle in the stereotaxic apparatus and controlling the injection rate using an electronic pump instead of injecting by hand. Furthermore, the stopper on the needle allows improved control of the puncture distance. The placement of the animal in the stereotaxic apparatus is crucial to control the puncture direction. If the animal is not placed correctly, the injection coordinates will not correspond to the area of interest. Importantly, the electronic pump connected to the Neuros syringe maintains a constant desired injection rate. A slow rate of infusion is necessary to reduce the risk of damage to the brain caused by the high pressure of rapid injection. The control of puncture distance is also important. If the needle tip enters the cerebellum, CSF tracer will be retained in the tissue and fail to distribute throughout the subarachnoid space. Damage to the medulla is frequently fatal, whereas cerebellar damage from chronic injection can result in prostration and general animal behavior abnormalities.

Compared to more invasive procedures, such as lumbar puncture or cannula insertion, our modified intrathecal injection method does not require surgery. Surgery often results in operative damages to the skull and brain parenchyma. Our approach entails transient puncture of the dural membrane, such that intracranial pressure is only transiently perturbed and is quickly restored. Cannula insertion into the cisterna magna is often associated with nervous tissue damage during insertion and scarring due to the material and size of the cannula. The complex methods to fix the cannula by drilling holes into the occipital bone also cause the surgical procedure to be more complex and time-consuming [15, 16]. However, with practice, a single intrathecal injection procedure as described here can be completed in 10 min, allowing treatment of larger cohorts of animals. Although intracisternal infusions have potential for iatrogenic complications, including traumatic tissue injury, we did not observe intracranial hypertension resulting from CSF tracer injection at the cisterna magna at an infusion rate of 3 μL/min. However, some animals did not survive when the infusion rate was increased to 5 μL/min (data not shown). Based upon these findings, we further performed intracisternal aCSF infusion at 3 μL/min for 3 consecutive days each week for a total of 2 weeks. All animals ($n = 6$) survived, and animal weight was not significantly impacted by the chronic infusions. In addition, using this approach, animals do not need to be single housed after injection, thus minimizing confounding factors due to the effects of stress. Overall, this approach circumvents difficulties associated with delivery of chemotherapeutic agents to the central nervous system. Intrathecal delivery via the cisterna magna could therefore be routinely employed to administer therapeutic agents to treat metastatic brain tumors in preclinical mouse models. It is noted that direct injection into the CSF may result in limited drug penetration to the brain parenchyma due to the slow diffusion rates. Intrathecal drug delivery may therefore be limited in its capacity to treat brain tumors in the parenchyma.

In conclusion, intrathecal injection via cisternal magna is a fast, easy, and reliable procedure to deliver drugs that could not otherwise cross the blood-brain barrier directly to the CSF compartment. Our noninvasive approach has promising clinical potential to deliver long-term therapy against leptomeningeal metastatic disease arising from brain tumors and other types of cancers.

Acknowledgments

This research was supported by a grant from the US National Cancer Institute (R01-CA241192) and an award from The Matthew Larson Foundation (MLF).

Conflict of Interest The authors have no conflicts of interest to declare that are relevant to the content of this chapter.

References

1. Donovan LK et al (2020) Locoregional delivery of CAR T cells to the cerebrospinal fluid for treatment of metastatic medulloblastoma and ependymoma. Nat Med 26(5):720–731
2. Medulloblastoma circulating tumor cells form leptomeningeal metastases. Cancer Discov (2018) 8(4):383
3. Garzia L et al (2018) A Hematogenous route for Medulloblastoma leptomeningeal metastases. Cell 172(5):1050–1062 e14
4. Znidaric T et al (2019) Breast cancer patients with brain metastases or leptomeningeal disease: 10-year results of a national cohort with validation of prognostic indexes. Breast J 25(6):1117–1125
5. Waki F et al (2009) Prognostic factors and clinical outcomes in patients with leptomeningeal metastasis from solid tumors. J Neuro-Oncol 93(2):205–212
6. Boire A et al (2017) Complement component 3 adapts the cerebrospinal fluid for leptomeningeal metastasis. Cell 168(6):1101–1113. e13
7. Glitza IC et al (2020) Leptomeningeal disease in melanoma patients: an update to treatment, challenges, and future directions. Pigment Cell Melanoma Res 33(4):527–541
8. Le Rhun E, Taillibert S, Chamberlain MC (2013) Carcinomatous meningitis:

leptomeningeal metastases in solid tumors. Surg Neurol Int 4(Suppl 4):S265–S288
9. Nayar G et al (2017) Leptomeningeal disease: current diagnostic and therapeutic strategies. Oncotarget 8(42):73312–73328
10. Leal T et al (2011) Leptomeningeal metastasis: challenges in diagnosis and treatment. Curr Cancer Ther Rev 7(4):319–327
11. Pardridge WM (2016) CSF, blood-brain barrier, and brain drug delivery. Expert Opin Drug Deliv 13(7):963–975
12. Pardridge WM (2007) Blood-brain barrier delivery. Drug Discov Today 12(1–2):54–61
13. Pardridge WM (2012) Drug transport across the blood-brain barrier. J Cereb Blood Flow Metab 32(11):1959–1972
14. Le Rhun E et al (2017) Diagnosis and treatment patterns for patients with leptomeningeal metastasis from solid tumors across Europe. J Neuro-Oncol 133(2):419–427
15. Ineichen BV et al (2017) Direct, long-term intrathecal application of therapeutics to the rodent CNS. Nat Protoc 12(1):104–131
16. Xavier ALR et al (2018) Cannula implantation into the cisterna magna of rodents. J Vis Exp 135:57378
17. Yang L et al (2013) Evaluating glymphatic pathway function utilizing clinically relevant intrathecal infusion of CSF tracer. J Transl Med 11:107
18. Penn RD, York MM, Paice JA (1995) Catheter systems for intrathecal drug delivery. J Neurosurg 83(2):215–217
19. Mehta AM, Sonabend AM, Bruce JN (2017) Convection-enhanced delivery. Neurotherapeutics 14(2):358–371
20. Stine CA, Munson JM (2019) Convection-enhanced delivery: connection to and impact of interstitial fluid flow. Front Oncol 9:966
21. Kunwar S et al (2006) Safety of intraparenchymal convection-enhanced delivery of cintredekin besudotox in early-phase studies. Neurosurg Focus 20(4):E15
22. Lou Y, Rao Y, Feng Z (2018) Intrathecal pump implantation in the cisterna magna for treating intractable cancer pain. Case Rep Anesthesiol 2018:5287150
23. Schuler B et al (2009) Optimized surgical techniques and postoperative care improve survival rates and permit accurate telemetric recording in exercising mice. BMC Vet Res 5:28
24. Packer RJ, Rood BR, MacDonald TJ (2003) Medulloblastoma: present concepts of stratification into risk groups. Pediatr Neurosurg 39(2):60–67
25. Northcott PA et al (2019) Medulloblastoma. Nat Rev Dis Primers 5(1):11
26. Wu X et al (2012) Clonal selection drives genetic divergence of metastatic medulloblastoma. Nature 482(7386):529–533
27. Ramaswamy V et al (2016) Medulloblastoma subgroup-specific outcomes in irradiated children: who are the true high-risk patients? Neuro-Oncology 18(2):291–297
28. Ramaswamy V et al (2013) Recurrence patterns across medulloblastoma subgroups: an integrated clinical and molecular analysis. Lancet Oncol 14(12):1200–1207
29. Garzia L et al (2018) A hematogenous route for medulloblastoma leptomeningeal metastases. Cell 173(6):1549
30. Iliff JJ et al (2013) Cerebral arterial pulsation drives paravascular CSF-interstitial fluid exchange in the murine brain. J Neurosci 33(46):18190–18199
31. Iliff JJ et al (2012) A paravascular pathway facilitates CSF flow through the brain parenchyma and the clearance of interstitial solutes, including amyloid beta. Sci Transl Med 4(147): 147ra111
32. Lipinski CA (2004) Lead- and drug-like compounds: the rule-of-five revolution. Drug Discov Today Technol 1(4):337–341
33. Leeson PD, Springthorpe B (2007) The influence of drug-like concepts on decision-making in medicinal chemistry. Nat Rev Drug Discov 6(11):881–890
34. Chen Y et al (2013) Novel modified method for injection into the cerebrospinal fluid via the cerebellomedullary cistern in mice. Acta Neurobiol Exp (Wars) 73(2):304–311
35. Reijneveld JC, Taphoorn MJ, Voest EE (1999) A simple mouse model for leptomeningeal metastases and repeated intrathecal therapy. J Neuro-Oncol 42(2):137–142

Chapter 10

In Vivo Mouse Models of *Myc*-Driven Medulloblastoma

Wanchen Wang and Daisuke Kawauchi

Abstract

Group 3 medulloblastoma (MB_{Grp3}), predominantly affecting young children, is often associated with *MYC* amplification and/or overexpression, frequent metastasis, and poor prognosis. To elucidate the mechanisms underlying phenomena that enhance the malignancy of such cancers, it is imperative to use mouse models that reflect the characteristics of human Group 3 tumors. In this chapter, we demonstrate highly aggressive MB_{Grp3} mouse models by transducing cerebellar progenitor cells with *Myc* and a dominant-negative mutant form of *TP53*.

Key words Medulloblastoma, Mouse model, Brain tumor, MYC, In utero electroporation

1 Introduction

Medulloblastoma (MB) is one of the most frequently occurring malignant pediatric brain tumors that arise in the cerebellum. Recent extensive omics analyses have molecularly classified this disease into four major subgroups (Wingless/int (WNT), Sonic Hedgehog (SHH), Group 3, and Group 4), and each subgroup is thought to develop from a distinct cellular origin by different oncogenic signaling pathways [1–6]. Gene amplification of the oncogene *MYC* is observed in various tumors, including MBs, where it is identified in approximately 17% of patients within the subgroup known as Group 3 (MB_{Grp3}) [7]. Moreover, high *MYC* expression is also associated with a generally poor prognosis for patients. Indeed, the current WHO classification of brain tumors highlights *MYC* as a diagnostic gene for non-WNT/SHH MBs.

To explore effective treatment strategies for high-*MYC* MB_{Grp3}, the development of mouse models that mimic this genetic mutation is essential and could be a powerful tool. Currently, a representative model is the patient-derived xenograft (PDX) model, which is widely used in preclinical studies. However, the cranial injection of developed human tumor cells into adult

Michael Goldstein (ed.), *Translational Research Methods for Pediatric Brain Tumors*, Neuromethods, vol. 226,
https://doi.org/10.1007/978-1-0716-4754-7_10,

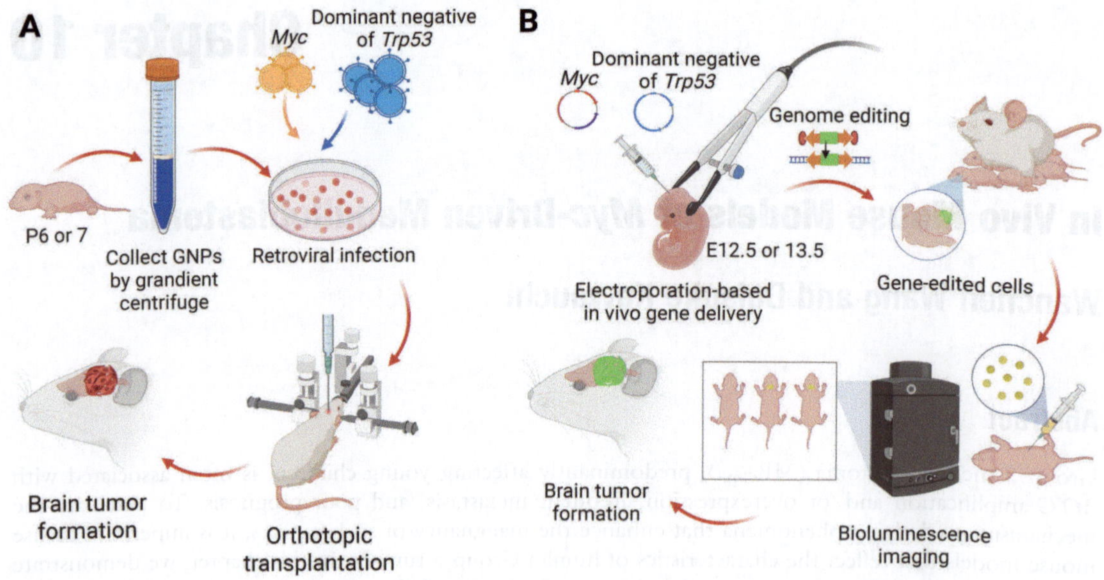

Fig. 1 Schematic diagrams of model generation strategies. **(A)** Illustration of MB_{Grp3} allograft model. **(B)** Illustration of MB_{Grp3} IUE model

immunocompromised mice has raised concerns regarding the lack of an intact immune system and the absence of a proper tumor microenvironment. In contrast, we have generated mouse models that progress under physiological conditions, which have been reported in previous studies by conditional activation of *Myc* in vivo together with functional inactivation of TRP53 in the developing cerebellum [6]. These models can be utilized to investigate the mechanisms of multistep tumorigenesis during tumor progression from various perspectives just as we previously did with our SHH-subgroup medulloblastoma (MB_{SHH}) model [8]. In this chapter, we introduce two models: an allograft model developed by an ex vivo viral transduction on cerebellar progenitors of postnatal mice and another that undergoes transformation under physiological conditions from the embryonic stage(s) in mice by in utero electroporation (IUE)-based oncogene transduction (*see* Fig. 1).

2 Materials

2.1 Mice

Either nude or NOD-SCID gamma (NSG) mice are used as recipients for MB_{Grp3} allograft models. Alternatively, *Cre*-carrying transgenic mice are subjected to IUE with plasmids for conditional activation of *Myc* and inactivation of TRP53 to induce MB_{Grp3} under a physiological condition. All procedures should be approved by local committees for the animal welfare.

2.2 Plasmids

For retrovirus production, pMSCV-Myc-IRES-RFP (Addgene, #35395), pMSCV-DNp53-copGFF, pMD old gagpol, and pCAG4-Eco are used [9]. For IUE, pT2K loxP-RFP-loxP (LRL) Myc-IRES-Luc, pT2K DNp53-copGFP (Addgene, #109229), and pCAG T2TP are used [6].

2.3 Reagents

1. Dulbecco's Modified Eagle Medium (Gibco, 11965092).
2. Fetal bovine serum (FBS) (Gibco, 10270106).
3. FuGENE HD Transfection Reagent (Promega, E2312).
4. Opti-MEM (Gibco, 31985070).
5. 1× DPBS, no calcium, no magnesium (Gibco, 14190144).
6. 10× DPBS, no calcium, no magnesium (Gibco, 14200075).
7. 0.5 M EDTA, pH 8.0 (Gibco, 15575020).
8. Neuralbasal medium (Gibco, 21103049).
9. Basal Medium Eagle (Gibco, 21010046).
10. GlutaMAX (Gibco, 35050061).
11. 45% D-Glucose (Sigma, G8769).
12. SPITE medium supplement (Sigma, S5666).
13. Linoleic acid (Sigma, L9655).
14. Penicillin/Streptomycin (Gibco, 15140122).
15. B27 supplement (Gibco, 17504044).
16. Distilled Water (Gibco, 15230204).
17. Deoxyribonuclease (DNase) I (Sigma, DN25).
18. Trypsin (Worthington, LS003703).
19. Percoll (Sigma, P4937).
20. 0.4% Trypan blue (Gibco, 15250061).
21. Poly-D-Lysine (Millipore, A-003-E).
22. Ovomucoid (Worthington Biochemical, LS003085).
23. 30% BSA solution (Sigma, A9576).
24. NaOH (Fujifilm Wako, 194-09575).
25. Matrigel (Corning, 354230).
26. Dermal adhesive (Daiichi-Sankyo, 3921011-00570).
27. Tris (Fujifilm Wako, 201-06273).
28. Fast green (Fujifilm Wako, 061-00031).
29. D-Luciferin (GoldBio, LUCK-100).
30. Paraformaldehyde (PFA) (Fujifilm Wako, 162-16065).
31. OCT compound (Sakura, 4583).
32. Adhesive glass slide (Matsunami, MAS-01).
33. Prolong glass antifade mountant (Invitrogen, P36980).

34. Entellan new (Sigma, 1079610100).
35. Antibody for YAP1 (Cell Signaling Technology, 14074S).
36. Antibody for MKI67 (Invitrogen, 14-5698-82).

2.4 Equipment

1. Water bath (TAITEC, e-Thermobucket).
2. Microscope (Zeiss, AxioImager M2 and Stemi305).
3. 0.22 and 0.45 μm filters (Millipore, SLGPR33RS, SLHPR33RS).
4. Surgery tools (FST, 14072-10, 11103-09, 11052-10, 14028-10).
5. 15 mL tube (Falcon, 352096).
6. 10-cm culture dish (Corning, 430167).
7. Stereotactic frame (Narishige, SMM-100).
8. Micro motor drill (Shofu, Tas-35LX).
9. Syringe pump (KdScientific, LEGATO130).
10. Hamilton syringe (Hamilton, 80030).
11. Electroporator (Nepagene, CUY21).
12. Forceps-like electrodes (Nepagene, CUY650P5).
13. Glass capillary (Narishige, GD-1.2).
14. In vivo imaging system (Vilber, Newton 7.0).

3 Methods

3.1 MB_{Grp3} Allograft Model

3.1.1 Reagent Preparation for the MB_{Grp3} Allograft Model

1. Granule cell medium: Neurobasal medium supplemented with 1× GlutaMAX, 0.45% D-glucose, 100 U/mL Penicillin/Streptomycin, 1× B27 supplement, 1× SPITE medium supplement and 1× linoleic acid. Store at 4 °C.
2. DNase solution: Basal Medium Eagle supplemented with 0.05% DNase I and 0.45% D-glucose. Sterilize with a 0.22 μm-pore size filter and store at 4 °C.
3. Trypsin/DNase solution: 1× DPBS supplemented with 1% Trypsin, 0.1% DNase I, 0.024% NaOH and 0.15% MgSO4·7H2O. Add MgSO4·7H2O at the end to avoid precipitation. Sterilize with a 0.22 μm-pore size filter and store at −20 °C.
4. 4x DPBS-EDTA: 4× DPBS with 0.01 M EDTA. Use 10x DPBS, 0.5M EDTA and ddH_2O for making this solution. Filter sterilize and store at 4 °C.
5. 35% Percoll: For 50 mL, mix 12.5 mL 4× DPBS-EDTA, 17.5 mL Percoll and 20 mL sterilized distilled water. Store at 4 °C.

6. 60% Percoll: For 50 mL, mix 12.5 mL 4× DPBS-EDTA, 30 mL Percoll, 7.5 mL sterilized distilled water, and 0.15 mL 0.4% Trypan blue. Store at 4 °C.
7. Trypsin solution: 1× DPBS supplemented with 0.05% Trypsin, 0.1% DNase, 0.004% NaOH (adding 50 μL 1 N NaOH to 50 mL solution), and 0.15% MgSO4·7H2O. Add MgSO4·7H2O at the end to avoid precipitation. Sterilize with a 0.22 μm-pore size filter and store at −20 °C.
8. Ovomucoid solution: 1× DPBS supplemented with 0.04% ovomucoid, 0.2% BSA, and 0.004% NaOH (adding 50 μL 1 N NaOH to 50 mL solution). Sterilize with a 0.22 μm-pore size filter and store at −20 °C.

Day 1–Day 5

3.1.2 Virus Production for the MB_{Grp3} Allograft Model

Day 1

1. Plate 6 × 10^6 of HEK293T cells on a 10-cm culture dish with 10 mL of HEK293T culture medium (DMEM supplemented with 10% heat-inactivated FBS, 1x GlutaMAX and 100 U/mL Penicillin/Streptomycin) 1 day before transfection.

Day 2

1. For transfection, a mixture of 8 μg of retroviral plasmid vector (pMSCV-Myc-IRES-RFP or pMSCV DNp53-copGFP), 3 μg of the pMD old gagpol plasmid, 1.5 μg of the pCAG4-Eco plasmid, and 36 μL of FuGENE HD transfection reagent is prepared in 500 μL of Opti-MEM and incubated at room temperature for 20 min before adding to the cells.
2. The mixture is directly added dropwise to the cells in the medium, and the dish is gently shaken to mix.

Day 3

1. Twenty-four hours after incubation at 37 °C, replace the culture medium with 4 mL granule cell medium for viral collection.

Day 4

1. Harvest virus-containing media completely after 8–12 h incubation. Then, add 4 mL of a fresh granule cell medium again.

Day 5

1. Repeat **step 5** twice. As a result, a total of approximately 10-12 mL of virus-containing medium can be collected.

2. The collected viral supernatant is filtered through a 0.45 μm filter and stored at −80 °C for long-term storage.

Day 6

3.1.3 Granule Neuron Progenitor (GNP) Purification and Ex Vivo Infection for the MB_{Grp3} Allograft Model

1. All procedures are done under 4 °C unless specified.
2. Dissect cerebella from postnatal day (P) 7 mice with sterilized surgery tools. The cerebella are kept in cold DPBS on ice during dissection. The day of birth is defined as P0.
3. Remove meninges of the cerebella under the stereomicroscope.
4. Transfer the cerebella to a 15 mL polypropylene conical tube, remove DPBS as much as possible, and add 1 mL Trypsin/DNase solution. For each preparation, no more than six cerebella are recommended.
5. Incubate at 37 °C for 5 min.
6. Remove the trypsin/DNase solution and add 2 mL of DNase solution. Do not pipet before adding DNase solution.
7. Resuspend cells by carefully pipetting with P1000 tips until no tissue chunk is left. Avoid bubbles at this step.
8. Spin down for 5 min at 800×*g*, 4 °C.
9. During centrifuge, prepare Percoll gradient centrifuge tube. For this, first add 4 mL of 35% Percoll to a 15 mL tube. Next, add 4 mL of 60% Percoll to the bottom with P1000 pipette slowly. 60% Percoll will settle at the bottom. Since 60% Percoll is colored in blue, a clear boundary line between 35% and 60% should be observed (*see* Fig. 2). When adding 60% Percoll, raise the pipette gradually to make sure the tip is just under the boundary between 35% and 60% Percoll to minimize liquid disturbance.
10. After step 8, discard the supernatant carefully and resuspend the cells with 1 mL DNase solution after tapping the cell pellets. Then, add another 2 mL of DNase solution, resulting in a total of 3 mL of cell suspension in DNase solution.
11. Load the cell suspension into the Percoll centrifuge tube (from **step 9**) onto the wall slowly in order to minimize the disturbance of layered Percoll solution (*see* Fig. 2).
12. Spin down for 10 min at 2000×*g*, at 4 °C, breaks off.
13. After centrifuge, glial cells and Purkinje cells are settled between DNase solution and 35% Percoll. GNPs are settled between 35% and 60% Percoll solutions.
14. Collect all GNP-enriched population with a P1000 pipette and add directly to 13 mL cold DPBS in a 15 mL tube to dilute Percoll.

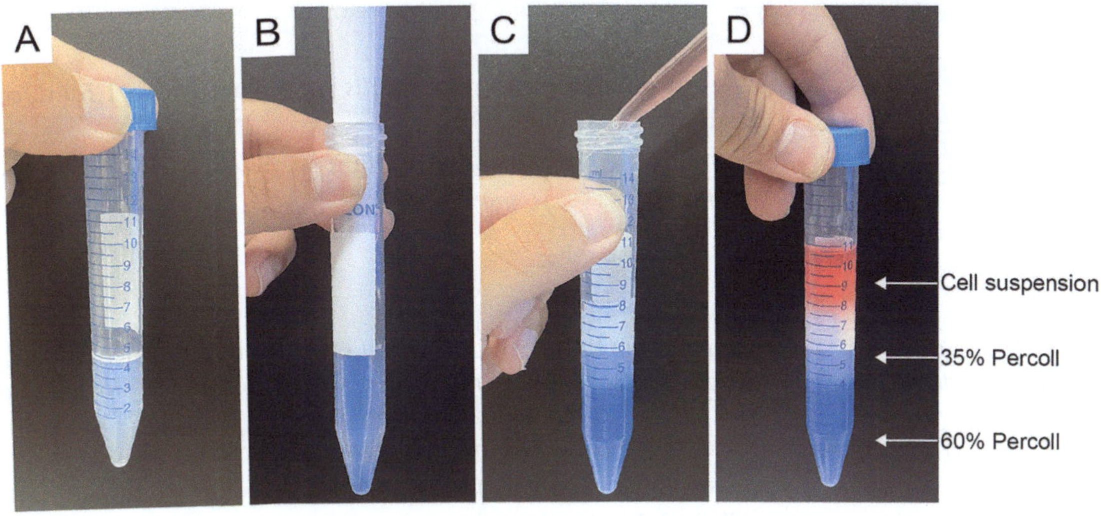

Fig. 2 Illustration of Percoll gradient centrifugation. (**A**) Add 4 mL 35% Percoll to the bottom of the tube. (**B**) Using a P1000 pipette, add 60% Percoll to the bottom of the tube slowly. (**C**) Load cell suspension onto the wall of the tube slowly. (**D**) Clear boundaries should be observed between different solutions in the gradient centrifuge tube

15. Spin down for 5 min at 800×*g*, 4 °C.
16. Discard supernatant carefully and resuspend cells with 5 mL granule cell medium after tapping.
17. Plate all cells onto a 35-mm dish. Incubate at 37 °C in 5% CO_2 for 1 h to allow fibroblasts to attach to the dish.
18. (Optional) Tap the plate gently and transfer non-attached cells to a new 35-mm dish, followed by additional incubation at 37 ° C in 5% CO_2 for an extra hour.
19. During **steps 17** and **18**, coat 6-well plates with 100 μg/mL poly-D-lysine for 30 min at room temperature. Then, remove poly-D-lysine and add 1:100 DPBS-diluted Matrigel working solution and incubate for 30 min at 37 °C in 5% CO_2.
20. Discard the Matrigel solution and rinse the coated-wells with DPBS for 3 times. Leave some DPBS in the wells until seeding the cells.
21. Tap the plate of **step 18** gently and transfer non-attached GNPs to a 15-mL conical tube. Count the cell number.
22. Plate five million of GNPs in 1 mL of granule cell medium per well.
23. The plated cells are infected three times at 2.5-h intervals with a mixture of 0.25 mL pMSCV-Myc-IRES-RFP and 0.25 mL pT2K DNp53-copGFP viruses at 37 °C in 5% CO_2, and then incubated overnight after the third infection.

Day 7

1. The culture media of infected cells are replaced with 1.5 mL of fresh granule cell medium.
2. The cells are incubated overnight at 37 °C in a 5% CO_2 atmosphere.

Day 8

3.1.4 Orthotopic Transplants for the Allograft Model

1. 48 h after infection, the cultured GNPs are treated with 500 μL of trypsin solution per well (in 6-well plates), followed by 500 μL of ovomucoid solution to stop the reaction, then the cells are collected in fresh granule cell media on ice.
2. Count cell number, make aliquots of 1.25×10^6 of the infected cells in 1.5 mL tubes (for one recipient mouse).
3. Spin them down for 5 min at 800×*g*, 4 °C.
4. Discard supernatant carefully and resuspend cells in a 1:1 mixture (5 μL in total) of granule cell medium and Matrigel on ice.
5. Anesthetize NSG mice aged over 8 weeks with 2% isoflurane and positioned on a stereotactic frame. Make a midline incision to expose the skull. Then drill a hole on the skull at the location of the cerebellum.
6. 4 μL of cell suspension (corresponding to 1.0×10^6 cells) is injected with a Hamilton syringe at a rate of 2 μL/min using a syringe pump at a depth of 3 mm.
7. Seal the head skin with a dermal adhesive.
8. After surgery, the mice are monitored daily. Sacrifice the mice once neurological symptoms (e.g., abnormal gaits, head dome, hunched posture) are observed. Of note, the symptoms can be observed from approximately 1 month onward after the operation. The tissue of interest (e.g., cerebellum and tumor) is used for further immunohistochemical analysis (GO TO Subheading 3.3) or is stored at −80 °C for molecular characterization, such as transcriptomics and proteomics.

3.2 IUE (Intrauterine Electroporation) Model

3.2.1 Animals for IUE Model

1. The male mice carrying the *Cre* gene driven by the regulatory element of either the *Atoh1* gene (JAX #011104) or the *Blbp* gene (NCI #01XM9) are bred with CD1 female mice. Subsequently, pregnant CD1 mice having embryonic day (E) 12.5 or E13.5 embryos are used for IUE operation. The day when vaginal plug is observed is defined as E0.5.

3.2.2 Plasmid Preparation for IUE Model

1. pT2K loxP-RFP-loxP (LRL) Myc-IRES-Luc, pT2K DNp53-copGFP (Addgene, #109229), and pCAG-T2TP are diluted with diluted TE (1 mM Tris-HCl, 0.1 mM EDTA, pH 8.0)

into 2 mg/mL, 2 mg/mL, and 0.5 mg/mL, respectively, at a final concentration. 0.01% fast green is added for visualization of the plasmid solution.

3.2.3 Electroporation-Based In Vivo Gene Transfer for IUE Model

1. IUE is performed as described in our previous study [10]. Briefly, 3–4 μL of the colored plasmid solution are injected into the fourth ventricle of the embryos through the uterine wall and the square electric pulses are delivered into the cerebellar primordium (30v for E12.5, 35v for E13.5, 50 ms duration, 950 ms interval, 5 pulses). Of note, *Cre*-expressing cells excise the LRL cassette by its DNA recombination activity, resulting in activation of *Myc* and *Luc* [6].

3.2.4 Luciferase Imaging

1. 5–7 days after birth (P5-P7), electroporated pups are subjected to intraperitoneal injection of 20 μL of D-luciferin (10 mg/mL in 0.9% NaCl).
2. Five minutes after injection, luciferase signal intensities are captured and measured as radiance (photons/s/cm^2/steradian) with NEWTON in vivo imaging system. Mice showing the strong signal intensity (more than 1×10^6 photons/s/cm^2/steradian) in the region of interest (e.g., mouse head) are returned to their mother and kept for the following analyses at defined stages or when the neurological symptoms are confirmed (GO TO Subheading 3.3).

3.3 Epifluorescent Microscopic Imaging

1. For analysis of tumor formation, the mice showing neurological symptoms are sacrificed and their cerebella are dissected out.
2. HE staining, DAB staining, and immunofluorescence staining can be performed according to conventional protocol.
3. Compared to MB_{SHH}, the following histological features can be observed in MB_{Grp3}: large anaplastic cells, high MKI67 positivity, and cytoplasmic localization of YAP1 instead of nuclear localization as seen in MB_{SHH} (*see* Fig. 3).

4 Notes

The development of the MYC-driven MB mouse model was carried out by enforced expression of MYC together with suppression or functional inhibition of TRP53 in cerebellar progenitor cells. However, in primary human cancers, loss-of-function mutations in *TP53* are not observed, leading to frequent discussions about the discrepancies between the mouse model and human cancers. Subsequently, *TP53* mutations were observed in samples of recurrent

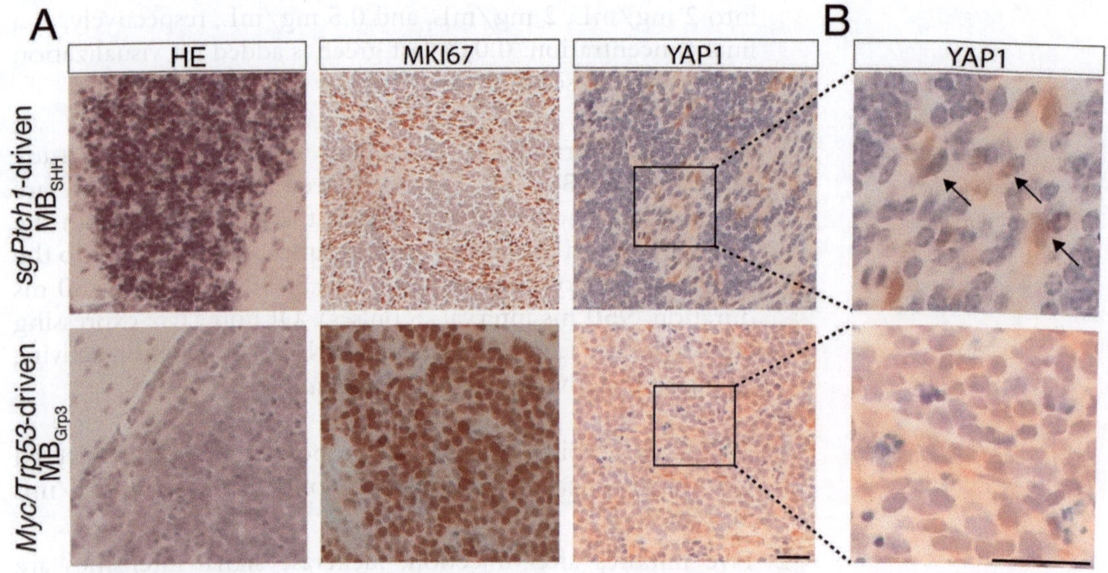

Fig. 3 Histology of murine medulloblastoma models. (**A**) Comparison of representative H&E staining and immunohistochemistry staining (IHC) with antibodies against MKI67 and YAP1 on *sgPtch1*-driven MB_{SHH} and *Myc/Trp53*-driven MB_{Grp3}. Scale bar, 25 μm. (**B**) Enlarged views of YAP1 IHC staining. Arrows indicate nuclear localization of YAP1 protein. Scale bar, 25 μm

MBs [11], and it was reported that GFI1 and GFI1B, known to attenuate the apoptotic function of *TP53* [12], are upregulated in human MYC-driven MBs through genomic rearrangements [13]. This has led to reconsideration of the existence of an indirect mechanism of *TP53* functional inhibition in MYC-driven MBs. Meanwhile, the origin of the cells remains inconclusive; although we and other researchers have used mouse genetics to investigate the tumorigenic potential of various cell types [6, 9, 14, 15], no definitive conclusions have been reached with validation using animal models. Recent omics-based studies on the human cerebellum have revealed that there are cell populations in humans that do not correspond to those in the murine cerebellum [3, 16]. It is believed that by exploring these correlations more closely, more refined murine MB models can be developed. Each model has its advantages and disadvantages: for example, while PDX models, which are widely used in preclinical research for human tumors, may be impacted by the lack of an immune system, IUE models maintain the immune system and brain microenvironment, but as they involve mouse tumors, differences in drug response due to interspecies differences cannot be ruled out. The allograft model allows for easier genetic manipulation without the need for specialized techniques and can be readily implemented in any laboratory. Thus, it is expected that these mouse models will be combined and advanced according to the research objectives.

Acknowledgments

We thank to Zhize Xiao for editing the manuscript and capturing images for Fig. 2, respectively.

Conflict of Interest The authors have no conflicts of interest to declare that are relevant to the content of this chapter.

References

1. Gibson P, Tong Y, Robinson G, Thompson MC, Currle DS, Eden C et al (2010) Subtypes of medulloblastoma have distinct developmental origins. Nature 468(7327):1095–1099
2. Visvanathan A, Saulnier O, Chen C, Haldipur P, Orisme W, Delaidelli A et al (2024) Early rhombic lip Protogenin(+ve) stem cells in a human-specific neurovascular niche initiate and maintain group 3 medulloblastoma. Cell 187:4733
3. Smith KS, Bihannic L, Gudenas BL, Haldipur P, Tao R, Gao Q et al (2022) Unified rhombic lip origins of group 3 and group 4 medulloblastoma. Nature 609(7929): 1012–1020
4. Yang ZJ, Ellis T, Markant SL, Read TA, Kessler JD, Bourboulas M et al (2008) Medulloblastoma can be initiated by deletion of patched in lineage-restricted progenitors or stem cells. Cancer Cell 14(2):135–145
5. Schuller U, Heine VM, Mao J, Kho AT, Dillon AK, Han YG et al (2008) Acquisition of granule neuron precursor identity is a critical determinant of progenitor cell competence to form Shh-induced medulloblastoma. Cancer Cell 14(2):123–134
6. Kawauchi D, Ogg RJ, Liu L, Shih DJH, Finkelstein D, Murphy BL et al (2017) Novel MYC-driven medulloblastoma models from multiple embryonic cerebellar cells. Oncogene 36(37):5231–5242
7. Northcott PA, Buchhalter I, Morrissy AS, Hovestadt V, Weischenfeldt J, Ehrenberger T et al (2017) The whole-genome landscape of medulloblastoma subtypes. Nature 547(7663):311–317
8. Shiraishi R, Cancila G, Kumegawa K, Torrejon J, Basili I, Bernardi F et al (2024) Cancer-specific epigenome identifies oncogenic hijacking by nuclear factor I family proteins for medulloblastoma progression. Dev Cell 59:2302
9. Kawauchi D, Robinson G, Uziel T, Gibson P, Rehg J, Gao CL et al (2012) A mouse model of the most aggressive subgroup of human medulloblastoma. Cancer Cell 21(2):168–180
10. Feng W, Herbst L, Lichter P, Pfister SM, Liu HK, Kawauchi D (2018) CRISPR-mediated loss of function analysis in cerebellar granule cells using in utero electroporation-based gene transfer. J Vis Exp 136:57311
11. Hill RM, Kuijper S, Lindsey JC, Petrie K, Schwalbe EC, Barker K et al (2015) Combined MYC and P53 defects emerge at medulloblastoma relapse and define rapidly progressive, therapeutically targetable disease. Cancer Cell 27(1):72–84
12. Khandanpour C, Phelan JD, Vassen L, Schutte J, Chen R, Horman SR et al (2013) Growth factor independence 1 antagonizes a p53-induced DNA damage response pathway in lymphoblastic leukemia. Cancer Cell 23(2): 200–214
13. Northcott PA, Lee C, Zichner T, Shuetz A, Erkek S, Kawauchi D et al. (2014) Enhancer hijacking activates GFI1 family oncogenes in medulloblastoma. Nature 511(7510):428–34. https://doi.org/10.1038/nature13379
14. Tao R, Murad N, Xu Z, Zhang P, Okonechnikov K, Kool M et al (2019) MYC drives group 3 medulloblastoma through transformation of Sox2(+) astrocyte progenitor cells. Cancer Res 79(8):1967–1980
15. Pei Y, Moore CE, Wang J, Tewari AK, Eroshkin A, Cho YJ et al (2012) An animal model of MYC-driven medulloblastoma. Cancer Cell 21(2):155–167
16. Visvanathan A, Saulnier O, Chen C, Haldipur P, Orisme W, Delaidelli A et al (2024) Early rhombic lip Protogenin(+ve) stem cells in a human-specific neurovascular niche initiate and maintain group 3 medulloblastoma. Cell 187(17):4733–50.e26

Chapter 11

In Vivo Model of Sonic Hedgehog Medulloblastoma by Postnatal Electroporation

Wanchen Wang and Daisuke Kawauchi

Abstract

Alongside the constitutive activation of Sonic hedgehog (SHH) pathway, SHH-subgroup medulloblastoma (MB_{SHH}) is characterized by additional mutations, including those in chromatin modifier genes, highlighting the significance of additional hits in its progression. To comprehensively assess the oncogenic potential of these mutations, a flexible and versatile animal model is essential. Here, we utilize CRISPR-mediated gene editing via electroporation-based gene transfer to knockout the genes in cerebellar granule neuron progenitors of neonatal mice. By knocking out the *Ptch1* gene at the neonatal stage, aberrant SHH activation is successfully induced, resulting in the formation of preneoplastic lesions and tumors. Of note, a simultaneous knockout of *Ptch1* and additional genes of interest can be achieved by incorporating multiple single-guide RNA cassettes into one single expression vector, enabling evaluation of the role of genetic mutations in the MB_{SHH}-prone context.

Key words Medulloblastoma, Mouse model, Brain tumor, Sonic hedgehog, Electroporation

1 Introduction

Medulloblastoma (MB) is one of the most prevalent pediatric brain tumors [1]. High-throughput genomics has categorized MBs into four distinct molecular subgroups: Wingless/int (WNT), Sonic Hedgehog (SHH), Group 3, and Group 4 [2–4]. Among these, the SHH subgroup MB (MB_{SHH}), which originates from cerebellar granule neuron precursors (GNPs) [5], accounts for about 30% of all MB cases and is marked by chronic activation of the SHH signaling pathway [5–7]. In MB_{SHH} patients, most have mutations in SHH pathway-related genes, including loss-of-function (LOF) mutations in *PTCH1* [8]. During typical cerebellar development, GNPs proliferate in the external granule layer (EGL) upon stimulation of SHH, secreted from Purkinje cells [9]. These cells then exit the cell cycle at appropriate stages to begin inward migration, eventually maturing in the internal granular layer (IGL). However,

Michael Goldstein (ed.), *Translational Research Methods for Pediatric Brain Tumors*, Neuromethods, vol. 226,
https://doi.org/10.1007/978-1-0716-4754-7_11,

GNPs lacking *Ptch1* show impaired cell cycle exit and migration, resulting in the formation of preneoplastic lesions on the surface of the developing cerebellum, starting from postnatal day (P) 21 [10]. Despite this, most preneoplastic cells (PNCs) ultimately differentiate into nonmalignant cells [11], highlighting the necessity for additional oncogenic events in tumor development [9].

For the treatment of MB_{SHH} patients, inhibition of abnormal SHH signaling activation has been considered a promising strategy. Despite remarkable progress in the development of FDA-approved SMO inhibitors, such as Vismodegib (GDC-0449) and Sonidegib (LDE225), their clinical use is limited due to severe side effects and the emergence of additional mutations that cause drug resistance [12–14], requiring further understanding of MB_{SHH} etiology for discovery of new therapeutic targets. To this end, recurrent mutations discovered by large-scale genome sequencing data on MBs [15] provided the opportunity to identify new oncogenic mechanisms, thus enhancing the efficacy of existing chemotherapy regimens.

In this study, we present flexible and versatile animal models for MB_{SHH} using CRISPR/Cas9-mediated gene editing of *Ptch1*, a negative regulator of SHH signaling in GNPs of neonatal cerebella. This method successfully induces pre-neoplastic lesions and tumors [16]. Additional mutations of MB_{SHH}-relevant genes are introduced by electroporating vectors carrying *spCas9* with specific single-guide RNAs (sgRNAs) alongside *sgPtch1*, enabling the investigation of specific mutations identified in human specimens within an MB_{SHH}-prone context on a mid-throughput scale [17].

2 Materials

2.1 Mice

In a *Ptch1* heterozygous genetic background, *Atoh1* expression persists during the transformation of GNPs into malignant cells [16–19], thus successfully accomplishing the genetic labeling of these transforming GNPs in vivo using EGFP fused to ATOH1 in mice (*see* Fig. 1). *Atoh1*^{EGFP} knock-in mice (JAX No.013593) are obtained from Jackson's laboratory [20].

Transforming GNP lineage is marked by EGFP (green) fused to ATOH1 in P7 GNPs (upper lane), P28 hyperplasia lesion (middle lane) and tumor (bottom lane). Scale bar, 100 μm. EGL, external granule layer; PNC, preneoplastic cell.

2.2 Plasmids

1. pX330-U6-Chimeric-BB-CBh-hSpCas9 Plasmid (Addgene #42230) [21].
2. pT2K-IRES-Luc [22].

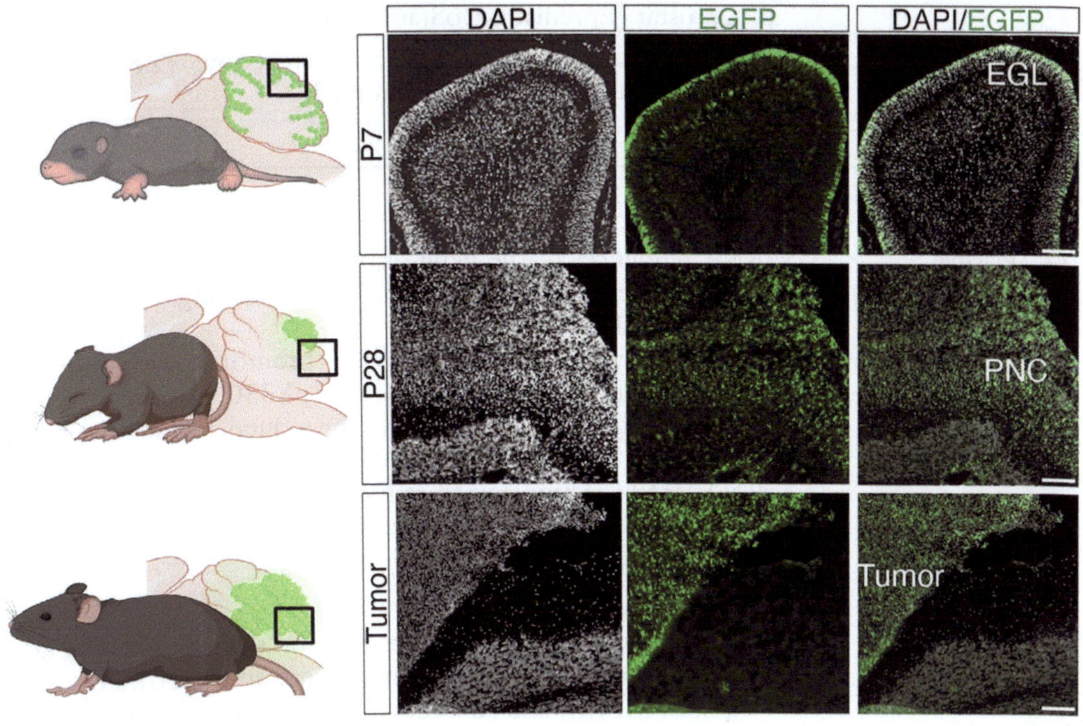

Fig. 1 Transformation if GNPs

2.3 Reagents

1. Fast green (Thermo, A16520.14).
2. D-Luciferin (GoldBio, LUCK-100).
3. ASAP Primer sets [Eurofins, 23].
4. Restriction enzyme BpiI (Thermo, FD1014).
5. Restriction enzyme NheI (Thermo, FD0974).
6. Restriction enzyme XbaI (Thermo, FD0684).
7. Quick Ligation Kit (NEB, M2200L).
8. 1× DPBS, no calcium, no magnesium (Gibco, 14190144).
9. OCT compound (Sakura Finetek, 4583).
10. Triton-X (Sigma, 9036-19-5).
11. Adhesive glass slide (Matsunami, MAS-01).
12. Prolong glass antifade mountant (Invitrogen, P36980).

2.4 Equipment

1. Electroporator (Nepagene, CUY21).
2. Aspirator tube (Sigma, A5177).
3. Forceps-like electrodes (Nepagene, CUY650P5).
4. Glass capillary (Narishige, GD-1.2).
5. In vivo imaging system (Vilber, Newton 7.0).

6. Cryostat (Epredia, CryoStar NX50).
7. Epifluorescence stereoscopic microscopy (Zeiss, Imager AxioM2, equipped with Apotome3).

3 Methods

3.1 Plasmid Construction

sgRNA sequences of the gene of interest (GOI) can be obtained from published papers or online tools, including CRISPOR.org [24] or CHOPCHOP [25].

The cassettes encoding the U6 promoter with individual sgRNAs are inserted tandemly into a pX330 plasmid (Addgene #42230) using ASAP cloning [22]. The sequence of sgRNA targeting *Ptch1* is CTGGCCGGAAAGCGCCGCTG . As a control sgRNA, *sgNfib* is reported to reduce the formation of hyperplasia lesions at P28 in our previous study [16].

Day 1

3.2 Electroporation

1. Electroporation-based gene transfer on the postnatal day (P) 5 is performed according to our previous study [26]. For induction of MB_{SHH} and subsequent in vivo screening, the pX330-sgRNA and pT2K IRES-Luc plasmids are diluted into 2 mg/mL and 0.5 mg/mL, respectively, as a final concentration. 0.01% fast green is added for visualization of the plasmid solution.
2. After anesthesia, a fusiform incision is made to expose the skull above the cerebellum.
3. Inject around 5 μL DNA plasmids into the primary fissure of the cerebellum with a glass capillary connected to an aspirator tube.
4. After injection of the plasmid solution into the area between the cerebellar surface and skull, square electric pulses (90v, 50 ms duration, 950 ms interval, 7 pulses) are delivered with forceps-like electrodes (*see* Fig. 2).
5. The operated mice are returned to their mothers and are nurtured by maternal care through breastfeeding until the time of analysis. Weaned animals that exhibit neurological symptoms (e.g., abnormal gaits, head dome, hunched posture) are sacrificed for further analysis, such as immunofluorescence staining.

Day 3

3.3 Luciferase Imaging

1. 2 days after P5 electroporation (P7), 20 μL of 10 mg/ml of D-luciferin intraperitoneally into electroporated pups is injected.

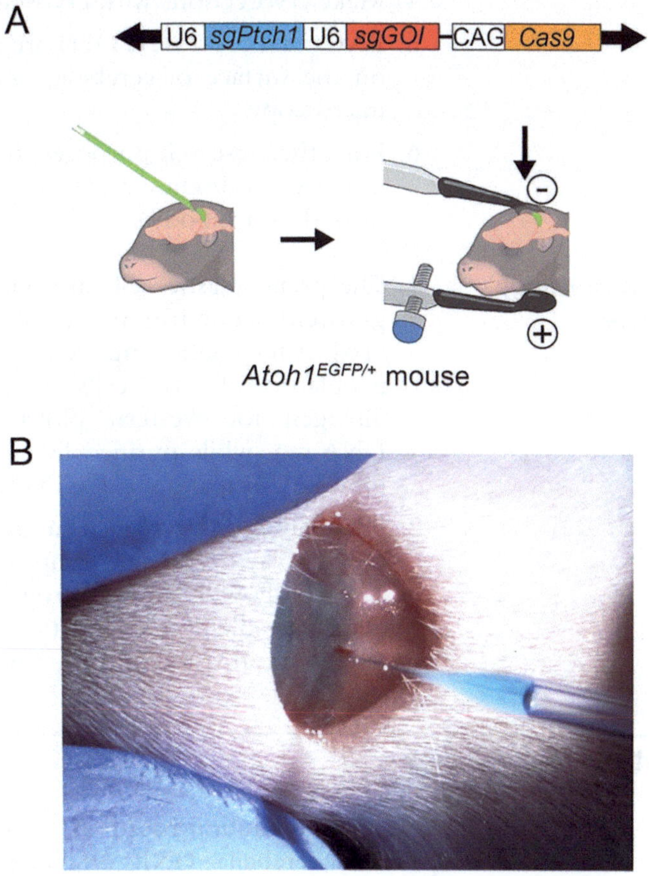

Fig. 2 In vivo electroporation on the postnatal cerebellum. (**A**) At P5, the plasmids encoding sgRNAs for *Ptch1* and gene of interest (GOI) and *luc* are electroporated into the cerebella of *Atoh1*^{EGFP} mice. The anode and cathode of the electrodes are placed on ventral and dorsal sides of the head. (**B**) The photo of the mouse cerebellum on surgery. The injected blue-colored plasmid solution is shown between the cerebellar surface and skull

2. Five minutes later, luciferase signal intensities are captured and measured as radiance (photons/s/cm^2/steradian) with NEWTON in vivo imaging system. Mice showing a strong signal intensity (more than 1×10^6 photons/s/cm^2/steradian) in the region of interest (mouse cranium) are kept for analysis.

3.4 Epifluorescence Microscopic Imaging

1. For analysis of hyperplasia formation, the operated mice are sacrificed at P28, and their cerebella are dissected out and fixed with 4% PFA/PBS at 4 °C overnight.
2. After fixation, the samples are transferred to 30% sucrose/PBS solution and further incubated at 4 °C overnight for cryoprotection.
3. Samples are embedded with OCT compound in a clear mold, yielding cryopreserved blocks. The blocks can be stored at − 80 °C until cryosectioning.

4. Make cryosections with cryostat at the thickness of 10 μm.
5. PNCs that express ATOH1 are confirmed by EGFP expression on the surface of cerebella under an epifluorescence stereo-microscopy.
6. For other research purposes, the sections can be stained with primary antibodies and corresponding fluorescence-labeled secondary antibodies.

3.5 Cellular Biology Experiments

1. The preneoplastic cells are purified by Percoll-based density gradient centrifugation, as described in Chapter 10 [16]. After confirming EGFP expression in the purified preneoplastic cells, the cells are either rapidly frozen in liquid nitrogen for western blotting, RNAseq, ChIP-seq, and DNA-seq, while purified live cells are immediately processed for ATAC-seq and CUT&Tag [16, 27].
2. For analysis on developed tumors, the mice are sacrificed when they display neurologic symptoms. Tumor chunks can be frozen rapidly with liquid nitrogen or purified by Percoll-based density gradient centrifugation, followed by subsequent analyses (e.g., in vitro culture), if needed.

4 Notes

Our electroporation-based in vivo gene transfer technique for postnatal cerebellar GNPs offers a flexible and versatile model for MB_{SHH}, providing a powerful tool for functional screening of potential oncogenic mutations and molecules involved in MB_{SHH} formation. Given that recurrent mutations have been identified in previous deep-sequencing studies on human primary MBs, this approach can expedite the identification of oncogenic mutations that confer mutation-specific traits, such as cell proliferation signaling and biomarkers, to tumor cells.

Acknowledgments

We thank Drs. K. Yamamoto and R. Shiraishi (National Center of Neurology and Psychiatry) for the helpful discussion of animal experiments. This work is generously supported by the grants from the Japan Agency for Medical Research and Development, AMED (JP23ama221129 to DK), the Brain Science Foundation (to DK), the NAITO Foundation (to DK), and the JST SPRING program (JPMJSP2120 to WW).

Conflict of Interest The authors have no conflicts of interest to declare that are relevant to the content of this chapter.

References

1. Ostrom QT, Gittleman H, Truitt G, Boscia A, Kruchko C, Barnholtz-Sloan JS (2018) CBTRUS statistical report: primary brain and other central nervous system tumors diagnosed in the United States in 2011–2015. Neuro-Oncology 20:1–86
2. Northcott PA, Jones DTW, Kool M, Robinson GW, Gilbertson RJ, Cho YJ et al (2012) Medulloblastomics: the end of the beginning. Nat Rev Cancer 12(12):818–834
3. Taylor MD, Northcott PA, Korshunov A, Remke M, Cho YJ, Clifford SC et al (2012) Molecular subgroups of medulloblastoma: the current consensus. Acta Neuropathol 123(4): 465–472
4. Kumar R, Liu APY, Northcott PA (2020) Medulloblastoma genomics in the modern molecular era. Brain Pathol 30(3):679–690
5. Yang ZJ, Ellis T, Markant SL, Read TA, Kessler JD, Bourboulas M et al (2008) Medulloblastoma can be initiated by deletion of patched in lineage-restricted progenitors or stem cells. Cancer Cell 14(2):135–145
6. Schüller U, Heine VM, Mao J, Kho AT, Dillon AK, Han YG et al (2008) Acquisition of granule neuron precursor identity is a critical determinant of progenitor cell competence to form Shh-induced medulloblastoma. Cancer Cell 14(2):123–134
7. Cavalli FMG, Remke M, Rampasek L, Peacock J, Shih DJH, Luu B et al (2017) Intertumoral heterogeneity within medulloblastoma subgroups. Cancer Cell 31(6):737–54 e6
8. Kool M, Jones DT, Jager N, Northcott PA, Pugh TJ, Hovestadt V et al (2014) Genome sequencing of SHH medulloblastoma predicts genotype-related response to smoothened inhibition. Cancer Cell 25(3):393–405
9. Wang W, Shiraishi R, Kawauchi D (2022) Sonic hedgehog signaling in cerebellar development and cancer. Front Cell Dev Biol 10: 864035
10. Oliver TG, Read TA, Kessler JD, Mehmeti A, Wells JF, Huynh TT et al (2005) Loss of patched and disruption of granule cell development in a pre-neoplastic stage of medulloblastoma. Development 132(10):2425–2439
11. Kessler JD, Hasegawa H, Brun SN, Emmenegger BA, Yang ZJ, Dutton JW et al (2009) N-myc alters the fate of preneoplastic cells in a mouse model of medulloblastoma. Genes Dev 23(2):157–170
12. Rudin CM, Hann CL, Laterra J, Yauch RL, Callahan CA, Fu L et al (2009) Brief report: treatment of medulloblastoma with hedgehog pathway inhibitor GDC-0449. New Engl J Med 361(12):1173–1178
13. Yauch RL, Dijkgraaf GJ, Alicke B, Januario T, Ahn CP, Holcomb T et al (2009) Smoothened mutation confers resistance to a hedgehog pathway inhibitor in medulloblastoma. Science 326(5952):572–574
14. Li Y, Song Q, Day BW (2019) Phase I and phase II sonidegib and vismodegib clinical trials for the treatment of paediatric and adult MB patients: a systemic review and meta-analysis. Acta Neuropathol Commun 7(1):123
15. Northcott PA, Buchhalter I, Morrissy AS, Hovestadt V, Weischenfeldt J, Ehrenberger T et al (2017) The whole-genome landscape of medulloblastoma subtypes. Nature 547(7663):311–317
16. Shiraishi R, Cancila G, Kumegawa K, Torrejon J, Basili I, Bernardi F et al (2024) Cancer-specific epigenome identifies oncogenic hijacking by nuclear factor I family proteins for medulloblastoma progression. Dev Cell 59:2302
17. Wang W, Kumegawan K, Chapman O, Shiraishi R, Xiao Z, Okonechnikov K, Pfister SM, Feng W, Uesaka N, Hoshino M, Takahashi S, Korshunov A, Chavez L, Maruyama R, Kawauchi D. Chromatin Modification Abnormalities by CHD7 and KMT2C Loss Promote Medulloblastoma Progression. Cell Rep 2025:44(5):115673
18. Ayrault O, Zhao HT, Zindy F, Qu CX, Sherr CJ, Roussel MF (2010) Atoh1 inhibits neuronal differentiation and collaborates with Gli1 to generate medulloblastoma-initiating cells. Cancer Res 70(13):5618–5627
19. Lee Y, Miller HL, Jensen P, Hernan R, Connelly M, Wetmore C et al (2003) A molecular fingerprint for medulloblastoma. Cancer Res 63(17):5428–5437
20. Rose MF, Ren J, Ahmad KA, Chao HT, Klisch TJ, Flora A et al (2009) Math1 is essential for the development of hindbrain neurons critical for perinatal breathing. Neuron 64(3): 341–354
21. Cong L, Ran FA, Cox D, Lin S, Barretto R, Habib N, Hsu PD, Wu X, Jiang W, Marraffini LA, Zhang F (2013) Multiplex genome engineering using CRISPR/Cas systems. Science 339(6121):819-23. https://doi.org/10.1126/science.1231143. PubMed 23287718
22. Kawauchi D, Ogg RJ, Liu L, Shih DJH, Finkelstein D, Murphy BL et al (2017) Novel MYC-driven medulloblastoma models from

multiple embryonic cerebellar cells. Oncogene 36(37):5231

23. Zuckermann M, Hlevnjak M, Yazdanparast H, Zapatka M, Jones DTW, Lichter P et al (2018) A novel cloning strategy for one-step assembly of multiplex CRISPR vectors. Sci Rep 8(1): 17499
24. Concordet JP, Haeussler M (2018) CRISPOR: intuitive guide selection for CRISPR/Cas9 genome editing experiments and screens. Nucleic Acids Res 46(W1):W242–W2W5
25. Labun K, Montague TG, Krause M, Torres Cleuren YN, Tjeldnes H, Valen E (2019) CHOPCHOP v3: expanding the CRISPR web toolbox beyond genome editing. Nucleic Acids Res 47(W1):W171–W1W4
26. Owa T, Taya S, Miyashita S, Yamashita M, Adachi T, Yamada K et al (2018) Meis1 coordinates cerebellar granule cell development by regulating Pax6 transcription, BMP signaling and Atoh1 degradation. J Neurosci 38(5): 1277–1294
27. Kutscher LM, Okonechnikov K, Batora NV, Clark J, Silva PBG, Vouri M et al (2020) Functional loss of a noncanonical BCOR-PRC1.1 complex accelerates SHH-driven medulloblastoma formation. Genes Dev 34(17–18): 1161–1176

Chapter 12

Techniques for Syngeneic Allograft Injections in Mice for Establishment of Pediatric Brain Tumor Models

Elton VanNoy, Abigail Haffey, Kaleem Coleman, Serge Yaacoub, Lynne El Ghorayeb, Khatereh Khorsandi, and Dalia Haydar

Abstract

Pediatric central nervous system (CNS) tumors pose a significant challenge in pediatric oncology, as they are the leading cause of cancer-related mortality among children. To address this critical issue, the development of reliable animal models that accurately replicate pediatric tumorigenesis is essential for advancing therapeutic strategies. This chapter outlines refined techniques for stereotactic intracranial injection models that reproduce exact anatomical locations and crucial immune characteristics of a tumor, thereby enabling robust assessments of immunotherapy efficacy in age-appropriate murine systems. Moreover, these models are vital tools for the preclinical evaluation of therapeutic interventions targeting the tumor immune microenvironment. We provide a detailed examination of the implantation modalities used for various tumor types, including injections into the cortex, cerebellum, fourth ventricle, and pons. The precision of these injections is imperative for accurately modeling tumor growth and the associated immune responses within these distinct brain regions. Ultimately, our comprehensive protocol serves as a methodological framework for preclinical studies of pediatric brain tumors, facilitating rigorous evaluations of therapeutic approaches in orthotopic murine models.

Keywords Pediatric brain cancer, Immunocompetent murine models, Stereotactic injection, Tumor anatomical localization

1 Introduction

Central nervous system (CNS) tumors are the most common solid tumors in pediatric patients and the second most common malignancy in children overall [1]. Glioma, ependymoma, and medulloblastoma are the most prevalent types of pediatric brain cancer, collectively accounting for 60–70% of all pediatric CNS tumors [1, 2]. Glioma is the most frequently occurring type of brain tumor in children, originating from glial cells and often occurring in the cortex [3]. It is classified into low-grade glioma (LGG) and

Michael Goldstein (ed.), *Translational Research Methods for Pediatric Brain Tumors*, Neuromethods, vol. 226,
https://doi.org/10.1007/978-1-0716-4754-7_12,

high-grade glioma (HGG) based on the tumor's histological features, aggressiveness, and growth rate. Diffuse midline glioma (DMG), a subtype of HGG, is associated with the highest mortality rate among pediatric brain tumors due to its location in the pons [4, 5]. Ependymomas arise from radial glial cells lining the ventricular system of the brain, often in the fourth ventricle. Due to proximity to the spinal cord, tumor growth often induces hydrocephalus and is difficult to surgically resect [6]. Targeting medulloblastoma presents unique challenges due to its significant heterogeneity, which includes four distinct molecular subtypes: WNT, Sonic Hedgehog, group 3, and group 4. Each subtype originates in the medulla and expands into the cerebellum, leading to compression of critical brainstem structures [7].

While the overall survival rate for pediatric brain cancer has increased over time, the prognosis varies significantly among different tumor types [2, 5]. High-risk subtypes continue to have a poor outlook, and central nervous system (CNS) tumors remain the leading cause of cancer-related deaths in children [8]. Additionally, despite significant advancements in pediatric cancer diagnostics and therapies over the past decade, treatment options remain limited and are frequently associated with severe side effects [1, 9]. Immunotherapy has shown significant success in treating cancers, particularly with the advent of FDA-approved chimeric antigen receptor (CAR) T cell therapies for hematologic malignancies [10]. CAR T cell immunotherapies have demonstrated safety in clinical trials against solid and CNS tumors but have yet to replicate the promising success observed in preclinical xenograft models [11]. To effectively apply immunotherapies to solid tumors, it is essential that in vivo models used in research and development accurately reflect key tumor features including the precise anatomic location and tumor immune microenvironment (TIME) of pediatric cancers [12]. Developing immunocompetent murine models that accurately represent the various subtypes of pediatric tumors is essential for advancing research and enhancing therapeutic strategies tailored to each subtype of pediatric CNS tumors. To establish reliable models of pediatric brain tumors, we have developed a syngeneic stereotactic injection methodology that not only mimics the precise anatomical localization of tumorigenesis but also utilizes immunocompetent and age-appropriate murine populations to accurately reflect immune system interactions. These tumor implantation techniques are robust, reliable, and relatively easy to implement within a reasonable timeframe, making them suitable for various preclinical studies. Consequently, this methodology serves as a valuable tool for assessing immunotherapies and understanding the tumor immune microenvironment (TIME) in pediatric tumors.

2 Materials for Stereotactic Injections

2.1 Equipment

1. Kopf Stereotactic frame.
2. Tabletop laboratory animal anesthesia system.
3. Surgical drill.
4. Warming pad.
5. Heating lamp.
6. Surgical microscope.
7. Tabletop bead sterilizer.

2.2 Tools

1. Insulin needle for IP injection.
2. Surgical scissors.
3. Micro-dissecting forceps.
4. Electric shaver.
5. Wound clip applier and remover.
6. Hamilton 10 μL Model 701 RN Syringe with Small Hub removable needle.

2.3 Supplies

1. Iodine wipes.
2. Alcohol wipes.
3. Cotton tipped applicator.
4. Eye lubricant.
5. Wound clips.
6. Isoflurane.
7. Pain reliever.
   - (a) Carprofen (Rimadyl).
     - (i) Dose: 5 mg/kg.

2.4 Mice

Mice should be 4–6 weeks old (*see* **Note 1**). C57BL/6 strain is a commonly used immunocompetent mouse line (*see* **Note 2**). A combination of male and female mice should be used to account for any sex-related difference.

2.5 Cell Lines

Several immunocompetent brain tumor models have been developed, and the select cell lines presented in Table 1 replicate specific genetic and pathological features of various pediatric brain tumors. These models have consistently demonstrated robust implantation in our hands and are highly relevant for immunotherapy-based research.

Table 1
Summary of murine orthotopic brain tumor models used for preclinical studies

Tumor type	Cell lines	Preparation	Location	Coordinates	Study	References
HGG	GL261 CT2A SB28	1×10^5 cells in 2 µl media/per injection 5×10^3 cells in 2µl media/per injection	Cortex	Lat: +2.0 Ant: +1.0 Vent: −4.0 **WRT Bregma**	Muller et al. (2012)	[13]
MB Group 3	2929 Myc driven (p53 −/−, p18 −/−) 2922 Myc driven (p53 −/−, p18 −/−) 1232 Myc driven (Gfi1 +/pl, p53 WT) (not published)	1×10^6 cells in 3µl/per injection	Cerebellum	Lat: 0.0 Ant: −2.5 to −3.0 Vent: −2.5 **WRT lambda**	Kawauchi et al. (2012) Vo et al. (2018) Vo et al. (2017)	[14] [15] [16]
MB SHH	63,718 MycN, p53-null	1×10^6 cells in 3µl/per injection	Cerebellum	Lat: 0.0 Ant: −2.5 to −3.0 Vent: −2.5 **WRT lambda**	Huang et al. (2010)	[17]
DMG	KPP 26-B (H3.3^{K27M}) 26-C (H3.1^{K27M}) 24-D (H3^{WT})	1×10^5 cells in 1µl/per injection 5×10^5 cells in 3µl/per injection	Pons	Lat: +1.0 Ant: −0.8 Vent: −5.5 **ERT lambda**	McNicholas et al. (2023) Marigil et al. (2017) Welby et al. (2019) Du Chatinier et al. (2022)	[18] [19] [20] [21]
Ependymoma (*see* **Note 3**)	MAF-928	2.5×10^5 cells in 3µl/per injection	4th ventricle	Lat: 0.0 Ant: −6.0 Vent: −4.3 **ERT Bregma**	Pierce et al. (2019)	[22]

Representative cell lines, preparation methods, anatomical injection sites, and stereotactic coordinates for intracranial implantation of high-grade glioma (HGG), medulloblastoma (MB), diffuse midline glioma (DMG), and ependymoma. Coordinates are with respect to (WRT) bregma or lambda, as noted. Corresponding references indicate studies describing or utilizing each model. Directional abbreviations are as follows: *lat* lateral, *ant* anterior, and *vent* ventral

3 Methods

3.1 Standard Injection Protocol

1. **Preparation of the Workstation**
   - (a) Clean the stereotactic frame and all working areas (*see* **Note 4**). Ensure the nose cone is secure, and anesthesia tubes are well fitted.
   - (b) Turn on the microscope and adjust positioning apparatus.
   - (c) Insert sterilized drill head intro surgical drill.
   - (d) Fill the vaporizer with isoflurane and begin oxygen flow at 2.5 L/min to prime the induction chamber.
     - (i) This flow rate is standard for mice to achieve a stable anesthetic concentration for induction.
   - (e) Prepare a recovery cage equipped with a heat source (e.g., heat lamp), ensuring access to food and water. Position food on the cage floor to accommodate postsurgical mice with limited mobility.
2. **Preparation of the Tumor Injection Syringe**
   - (a) Prepare the Hamilton syringe by washing with incomplete media.
   - (b) Load the cells into the syringe (*see* **Note 5**).
     - (i) Gently resuspend cells before taking volume.
     - (ii) Ensure there are no air bubbles loaded into syringe.
     - (iii) Load the Hamilton syringe for one injection at a time. Rinse with media and refill for each mouse.
   - (c) Attach the syringe to stereotactic frame.
     - (i) Ensure the volume gradients are visible.
3. **Preparation of Mice**
   - (a) Place the mouse in the induction chamber.
     - (i) Anesthetize at 2.5 isoflurane flow rate.
     - (ii) Use the toe pinch method to check the sedation level.
   - (b) Once anesthetized, carefully shave and clean the injection area.
     - (i) Ear tag the mouse for identification (optional).
     - (ii) Weigh the mouse and record initial body weight in grams.
   - (c) Inject carprofen solution (5 mg/mL) intraperitoneally to treat pain and inflammation with insulin syringe.
   - (d) Transfer the anesthetized mouse to the stereotaxic apparatus, on top of a preheated pad to maintain body heat; 37 °C is optimal.

(i) Following shaving and analgesic administration, the mouse may begin to regain consciousness. If signs of awakening are observed, return the animal to the induction chamber for re-anesthetization prior to positioning in the stereotaxic frame.

(e) Secure the mouse using ear bars positioned at 15 mm (or as anatomically appropriate), a bite bar set at 5 mm, and a nose cone to maintain anesthesia.

(i) Ensure the mouse's head and body is flat and its neck is not at a severe angle. If necessary, place extra support underneath the mouse's body to elevate it.

(ii) Confirm that anesthetic gas flow is properly directed to the nose cone following induction, and that the induction chamber remains connected if needed.

(f) Apply eye lubricant immediately.

(g) Monitor the mouse's breathing during surgery to gauge sedation level and general wellness.

(i) If breathing is restricted, loosen the nose cone.

(ii) If the mouse is regaining consciousness, remove the mouse from the stereotaxis and place it back in the induction chamber until adequate sedation is reached.

4. **Preparation of Injection Site**

(a) Use an alcohol wipe to remove loose fur from the head area.

(b) Apply iodine to clean the head.

(c) Using scissors and forceps, create an incision along the centerline of the mouse's head.

(d) Ensure the injection site is visible. Clear tissue using a cotton-tip applicator and surgical tools.

5. **Injection**

(a) Follow injection site-specific protocol in Subheadings 3.2, 3.3, 3.4, and 3.5 (see Fig. 1) (*see* **Note 6**).

6. **Mouse Recovery**

(a) Close the skin surrounding the wound on the skull using wound clips.

(i) Wound clips should be removed 72 h after surgery.

(b) Remove the nose cone, ear bars, and take the mouse off the stereotactic frame.

(c) Place the mouse in the recovery cage.

(i) Ensure the heat lamp is aimed toward half of the cage so overheated mice can move to the unheated portion of the cage.

(d) Monitor the mouse closely until it is awake and moving around the cage.

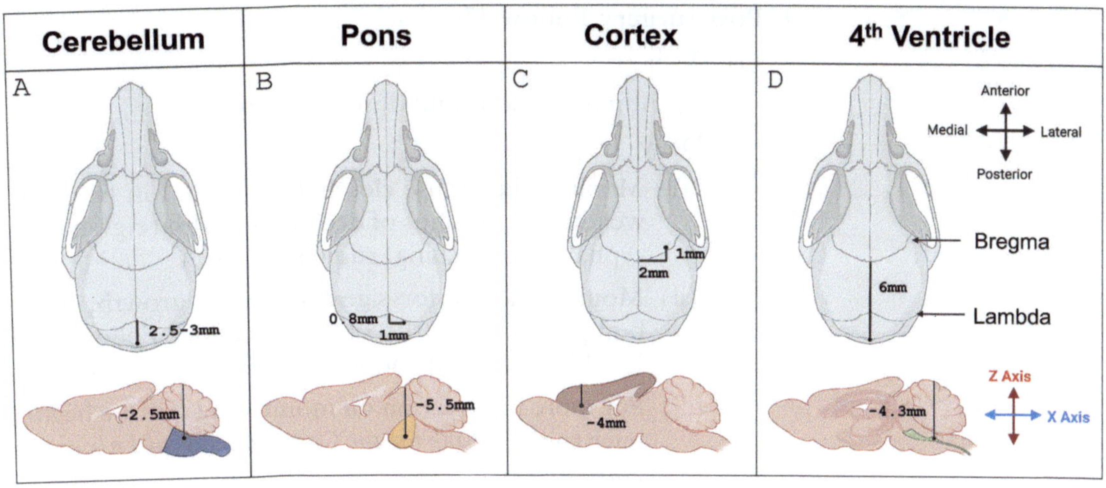

Fig. 1 Representative schematic illustrating stereotactic injection sites for the cerebellum (**a**), pons (**b**), cortex (**c**), and fourth ventricle (**d**) in relation to bregma and lambda sutures. The figure provides a general overview of target regions for injections, without anatomical precision. For proper orientation, refer to the directional legend included in the schematic and coordinates

7. **After Surgery**
   (a) Between mice:
      (i) Clean the needle with incomplete media.
      (ii) Clean surgical tools with alcohol between each mouse.
      (iii) Wipe drill tip with an alcohol wipe in between mice.
   (b) Between mouse cages:
      (i) Switch set surgical tools or sterilize with tabletop bead sterilizer.
      (ii) Change the injection needle.
8. **Cleanup**
   (a) Remove the drill tip from the drill.
   (b) Clean syringe needle with PBS, 70% ethanol, and then DI water.
   (c) When finished, clean the surgical area with 70% ethanol.
   (d) Autoclave drill tips and surgical tools.
      (i) Autoclaving may dull the drill tips; instead you may sterilize with the tabletop bead sterilizer.
   (e) Disassembled Hamilton syringes should be sterilized in 70% ethanol overnight, cleaned thoroughly with DI water, and then stored with the metal wire inserted to prevent clogging.

9. **Post-surgery Follow-Up**
   (a) 24 h
      (i) Administer a second dose of carprofen.
   (b) Daily
      (i) Monitor the mouse closely for changes in behavior, weight loss, or signs of declining health (e.g., hydrocephalus, scrunching, ruffled fur, etc.).
      (ii) Monitor closely for signs of tumor growth, like a bump forming at the site of injection. Follow your IACUC-approved endpoint criteria.
      (iii) For more information on monitoring murine health, refer to the *Guide for the Care and Use of Laboratory Animals* [23].
   (c) Weekly
      (i) Weigh mice and euthanize if weight loss is greater than 20% of initial recorded body weight.
         1. Weigh more frequently if weight loss is apparent.
   (d) As needed
      (i) Image mice to confirm tumor size and location using magnetic resonance imaging (MRI) (*see* Fig. 2) (*see* **Note** 7).

3.2 Cerebellum Injection Protocol

Cell Preparation Resuspend cells at a concentration of 1×10^6 cells per 3 μL in a 1:1 dilution of media and Matrigel. Injection volume is 3 μL per mouse.

1. **Follow Standard Injection Protocol for steps 1–4.**
2. Using the needle attached to the stereotaxic frame, position over **lambda.** Move the needle 2.5–3 mm back in the Z-axis (posteriorly). Mark this spot visually or by a small indentation and then use the handheld drill to create a burr hole.
   (a) When drilling, apply a small but firm amount of pressure.
   (b) If there is blood, gently wipe using a cotton-tip applicator.
3. Using the needle, position it directly at the top of the drilled hole. Lower the needle 2.5 mm deep into the brain. Let the needle sit for 1 min and then move 0.5 mm up.
4. Slowly inject 1 μL of cells and then wait for 1 min. Slowly move the needle up 0.5 mm.
5. Repeat **step 4** until the full 3 μL volume is injected.

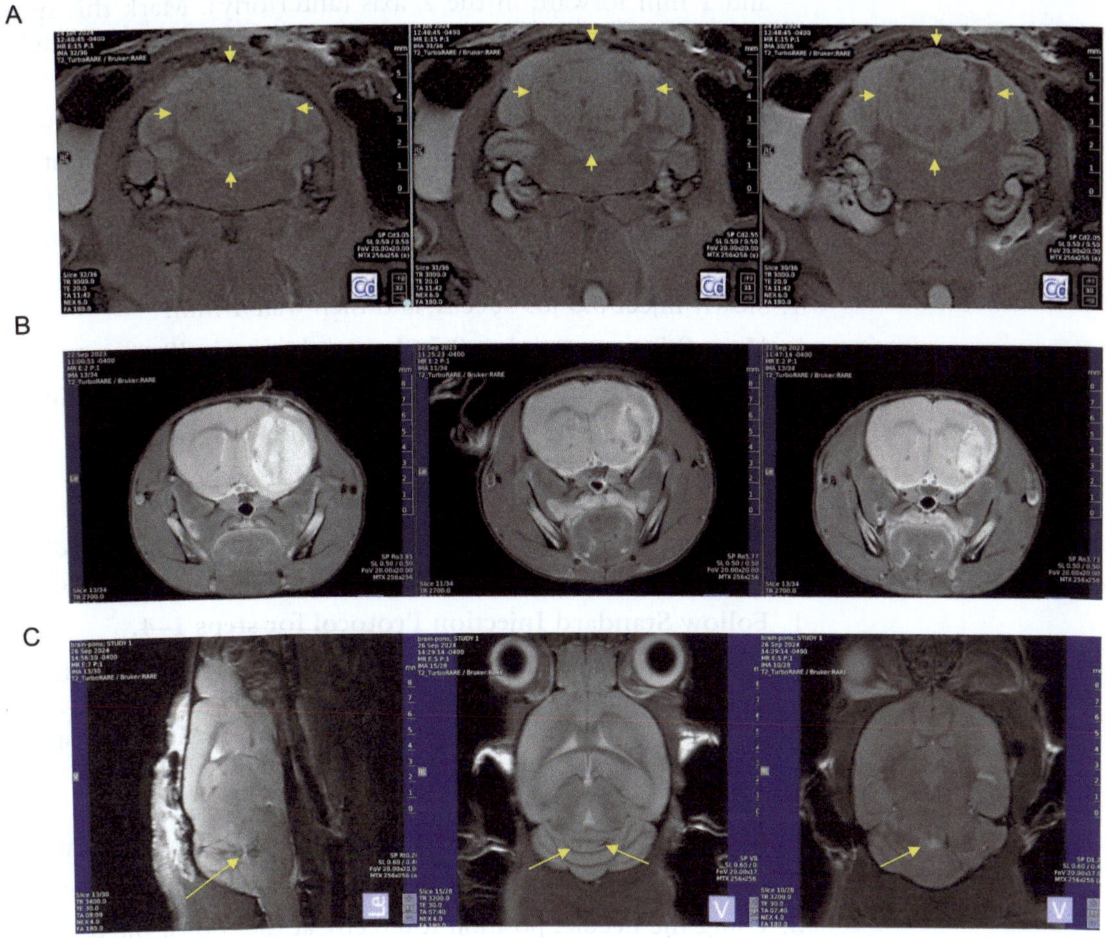

Fig. 2 Tracking tumor implantation by magnetic resonance imaging (MRI): (**a**) MRI images of C57BL/6 mice demonstrating successful implantation of group 3 medulloblastoma in the mouse cerebellum at day 15 post-implantation. (**b**) Representative MRI images of C57BL/6 mice showing successful implantation of high-grade glioma (HGG) in the cortex at day 20 post-tumor implantation. (**c**) MRI images of C57BL/6 mice demonstrating successful implantation of diffuse midline glioma (DMG) tumor (26B) in the pons at 14 days post-tumor implantation. All mice were 4–5 weeks of age at implantation and include a mix of male and female mice

6. Once the injection is completed, let the needle sit for 1 min. Then slowly move the needle up and out of the brain over 2 min.
7. **Follow Standard Injection Protocol for steps 6–9.**

3.3 Cortex Injection Protocol

Cell Preparation Resuspend cells at a concentration indicated in Table 1. Cell preparation varies depending on the cell type used. Injection volume is 2 μL per mouse.

1. **Follow Standard Injection Protocol for steps 1–4.**
2. Using the needle attached to the stereotaxic frame, position over **bregma**. Move the needle 2 mm to the right in the X-axis

and 1 mm forward in the Z-axis (anteriorly). Mark this spot visually or by a small indentation and then use the handheld drill to create a hole.

   (a) When drilling, apply a small but firm amount of pressure.
   (b) If there is blood, gently wipe using a cotton-tip applicator.

3. Using the needle, position it directly at the top of the drilled hole. Lower the needle 4 mm deep into the brain, and then lift the needle 1 mm up.
4. Slowly inject 0.5 μL of cells, and then wait 1 min.
5. Move 0.5 mm up, inject 0.5 μL, and then wait 30 s.
6. Repeat **step 5** until full volume is injected. Then wait 1 min.
7. Slowly raise the needle up and out of the brain over 2 min.
8. **Follow Standard Injection Protocol for steps 6–9.**

3.4 Pons Injection Protocol

Cell Preparation Resuspend cells at a concentration indicated in Table 1. Cell preparation varies depending on the cell type used.

1. **Follow Standard Injection Protocol for steps 1–4.**
2. Using the needle attached to the stereotaxic frame, position over **lambda**. Move the needle 1 mm to the right of the X-axis and 0.8 mm back in the Z-axis (posteriorly). Mark this spot visually or by a small indentation and then use the handheld drill to create a hole.
   (a) When drilling, apply a small but firm amount of pressure.
   (b) If there is blood, gently wipe using a cotton-tip applicator.
3. Using the needle, position it directly at the top of the drilled hole. Lower the needle 5.5 mm deep into the brain. Let the needle sit for 1 min, and then move 0.5 mm up.
4. Slowly inject 1 μL of cells, and then wait 1 min. Slowly move the needle up 0.5 mm.
5. Repeat **step 4** until the full volume is injected.
6. Once the injection is completed, let the needle sit for 1 min. Then, slowly move the needle up and out of the brain over 2 min.
7. **Follow Standard Injection Protocol for steps 6–9.**

3.5 Fourth Ventricle Injection Protocol

Cell Preparation Resuspend cells at a concentration of 2.5×10^5 cells per 3 μL in a 1:1 dilution of media. Injection volume is 3 μL per mouse.

1. **Follow Standard Injection Protocol for steps 1–4.**
2. Using the needle attached to the stereotaxic frame, position over the **bregma.** The needle remains on the midline and

moves 6 mm back on the Z-axis (posteriorly). Mark this spot visually or by a small indentation, and then use the handheld drill to create a hole.

(a) When drilling, apply a small but firm amount of pressure.

(b) If there is blood, gently wipe using a cotton-tip applicator.

3. Using the needle, position it directly at the top of the drilled hole. Lower the needle 4.3 mm deep into the brain. Let the needle sit for 1 min, and then move 0.5 mm up.
4. Slowly inject 1 μL of cells, and then wait 1 min. Slowly move the needle up 0.5 mm.
5. Repeat **step 4** until the full 3 μL volume is injected.
6. Once the injection is completed, let the needle sit for 1 min. Then slowly move the needle up and out of the brain over 2 min.
7. **Follow Standard Injection Protocol for steps 6–9.**

4 Notes

1. For the generation of pediatric murine models, it is important to consider the age correlation of mice to humans. The pediatric age range often includes years 0–19, indicating a maximum of 20-week-old mice for inclusion in pediatric studies. Mice progress through juvenile periods quickly as mouse age does not correlate linearly with the age of humans. For example, 4-week-old mouse is equivalent to a 1-year-old human, while a 5-week-old mouse is equivalent to a 5-year-old human [25].
2. If you are using mice delivered from a commercial vendor, allow at least 7 days for acclimation to their new environment.
3. The ependymoma models $1q^+$ PFA ependymoma patient-derived xenograft model (MAF) are not murine-derived. To be utilized in an immune-competent setting in mice, the mouse models should have a reconstituted immune system (humanized). This immune reconstitution involves full body irradiation of mouse models such as Balb/c Rag2-/- Il2rg-/- SirpaNOD (BRGS), followed by introduction of CD34+ hematopoietic stem cells derived from human umbilical cord blood. After 6 weeks, these mice are eligible for fourth ventricle injections. Additionally, ependymoma cells can be propagated and expanded in vivo in NSG mice via intracranial injections [22].
4. During the initial setup, create separate working areas for mouse preparation (i.e., shaving, applying iodine), surgery, and recovery.

5. Keep cells on ice for the duration of injections. Use cells within 4 h but ideally immediately following preparation.
6. Very slow injection of cells is necessary to prevent cells from flushing out of the cavity, resulting in tumor implantation below the skin.
7. Imaging is done primarily by magnetic resonance imaging (MRI) as opposed to in vivo imaging system (IVIS). Luciferase labeling of tumor models may cause potential immunological interactions affecting the tumor microenvironment [24].

Conflict of Interest

The authors have no conflicts of interest to declare that are relevant to the content of this chapter.

References

1. Hossain MJ, Xiao W, Tayeb M, Khan S (2021) Epidemiology and prognostic factors of pediatric brain tumor survival in the US: evidence from four decades of population data. Cancer Epidemiol 72:101942
2. Ostrom QT, Price M, Ryan K, Edelson J, Neff C, Cioffi G et al (2022) CBTRUS statistical report: Pediatric Brain Tumor Foundation childhood and adolescent primary brain and other central nervous system tumors diagnosed in the United States in 2014-2018. Neuro-Oncology 24(Suppl 3):1–38
3. Wang X, Zhou C, Wang L, Wang Y, Jiang T (2020) Motor cortex gliomas induces microstructural changes of large fiber tracts revealed by TBSS. Sci Rep 10(1):16900
4. Price G, Bouras A, Hambardzumyan D, Hadjipanayis CG (2021) Current knowledge on the immune microenvironment and emerging immunotherapies in diffuse midline glioma. EBioMedicine 69:103453
5. Li Z, Langhans SA (2021) In vivo and ex vivo pediatric brain tumor models: an overview. Front. Oncologia 11:620831
6. Toescu SM, Samarth G, Layard Horsfall H, Issitt R, Margetts B, Phipps KP et al (2021) Fourth ventricle tumors in children: complications and influence of surgical approach. J Neurosurg Pediatr 27(1):52–61
7. Fang FY, Rosenblum JS, Ho WS, Heiss JD (2022) New developments in the pathogenesis, therapeutic targeting, and treatment of pediatric Medulloblastoma. Cancers (Basel) 14(9): 2285
8. Udaka YT, Packer RJ (2018) Pediatric brain tumors. Neurol Clin 36(3):533–556
9. Nguyen T, Mueller S, Malbari F (2022) Review: neurological complications from therapies for pediatric brain tumors. Front Oncol 12:853034
10. Ahmad A, Uddin S, Steinhoff M (2020) CAR-T cell therapies: an overview of clinical studies supporting their approved use against acute lymphoblastic leukemia and large B-cell lymphomas. Int J Mol Sci 21(11):3906
11. Yan T, Zhu L, Chen J (2023) Current advances and challenges in CAR T-cell therapy for solid tumors: tumor-associated antigens and the tumor microenvironment. Exp Hematol Oncol 12(1):14
12. Haydar D, Ibañez-Vega J, Krenciute G (2021) T-cell immunotherapy for pediatric high-grade gliomas: new insights to overcoming therapeutic challenges. Front Oncol 11:718030
13. Muller F, Lee M, Muller F, Aquilanti E, Hu B, DePinho R (2012) Stereotactic orthotopic xenograft injections into the mouse brain. Protocol Exchange
14. Kawauchi D, Robinson G, Uziel T, Gibson P, Rehg J, Gao C et al (2012) A mouse model of the most aggressive subgroup of human medulloblastoma. Cancer Cell 21(2):168–180
15. Vo BT, Kwon JA, Li C, Finkelstein D, Xu B, Orr BA et al (2018) Mouse medulloblastoma driven by CRISPR activation of cellular Myc. Sci Rep 8(1):8733
16. Vo BT, Li C, Morgan MA, Theurillat I, Finkelstein D, Wright S et al (2017) Inactivation of Ezh2 upregulates Gfi1 and drives aggressive Myc-driven group 3 Medulloblastoma. Cell Rep 18(12):2907–2917

17. Huang X, Sarangi A, Ketova T, Litingtung Y, Cooper MK, Chiang C (2010) Intracranial orthotopic allografting of medulloblastoma cells in immunocompromised mice. J Vis Exp 44
18. McNicholas M, De Cola A, Bashardanesh Z et al (2023) A compendium of syngeneic, transplantable pediatric high-grade glioma models reveals subtype-specific therapeutic vulnerabilities. Cancer Discov 13:1592
19. Marigil M, Martinez-Velez N, Domínguez PD, Idoate MA, Xipell E, Patiño-García A et al (2017) Development of a DIPG orthotopic model in mice using an implantable guide-screw system. PLoS One 12(1):e0170501-e
20. Welby JP, Kaptzan T, Wohl A, Peterson TE, Raghunathan A, Brown DA et al (2019) Current murine models and new developments in H3K27M diffuse midline gliomas. Front Oncol 9:92
21. du Chatinier A, Meel MH, Das AI, Metselaar DS, Waranecki P, Bugiani M et al (2022) Generation of immunocompetent syngeneic allograft mouse models for pediatric diffuse midline glioma. Neurooncol Adv 4(1): vdac079-vdac
22. Pierce AM, Witt DA, Donson AM, Gilani A, Sanford B, Sill M et al (2019) Establishment of patient-derived orthotopic xenograft model of 1q+ posterior fossa group a ependymoma. Neuro-Oncology 21(12):1540–1551
23. Hubrecht R (2011) Guide for the care and use of laboratory animals, eighth edition 2011 the Committee for the Update of the guide for the care and use of laboratory animals (2011). Published by the National Research Council of the National Academies, Washington DC, USA. 219 pp paperback (ISBN 0-309-15400-6). Price US$19.95. Anim Welf 20(3): 455–456
24. Sanchez VE, Lynes JP, Walbridge S, Wang X, Edwards NA, Nwankwo AK et al (2020) GL261 luciferase-expressing cells elicit an anti-tumor immune response: an evaluation of murine glioma models. Sci Rep 10(1):11003
25. Wang S, Lai X, Deng Y, Song Y (2020) Correlation between mouse age and human age in anti-tumor research: significance and method establishment. Life Sci 242:117242

Chapter 13

Developing a Zebrafish Xenograft Model of Diffuse Midline Glioma

Kaixuan Wang, Gianna Graziano, Anneliese Ceisel, Huanhuan Xiao, Shreya Banerjee, Yuran Yu, Marina Venero Galanternik, Brant M. Weinstein, Charles G. Eberhart, Jeff S. Mumm, and Eric H. Raabe

Abstract

Diffuse midline glioma (DMG) is a highly aggressive brain tumor that predominantly affects children. Conventional treatments such as radiation therapy can control progression for a time, but DMG kills nearly 100 percent of patients. Although murine models have provided critical insights into the biology of DMG, they are not suitable for high-throughput screening to identify and profile novel therapies due to technical challenges, ethical considerations, and high cost. Zebrafish (*Danio rerio*) is an established vertebrate model for large-scale drug screening, and zebrafish have demonstrated the ability to replicate the key biological and pathological aspects of human malignancies.

Here, we developed a novel method for transplanting human DMG cells into large numbers of zebrafish embryos to speed the assessment of anti-tumor drug efficacy in vivo and thereby facilitate the development of novel therapeutics for clinical translation. We transplanted red fluorescent protein (RFP)-labeled, patient-derived DMG cell lines into zebrafish blastulas. Remarkably, many DMG cells migrate into the developing brain and are present in the midline of the brain 24 h after blastula injection. Tumor cell burden was monitored by measuring RFP fluorescence intensity changes over time. Time-course images of transplanted tumor cell volumes were acquired, and the interactions between transplanted DMG cells and microglial cells were further analyzed using Imaris software. We have developed a robust and rapid transplantation protocol to establish a zebrafish xenograft model of DMG. Unlike existing orthotopic zebrafish models, our protocol eliminates the need for technically demanding intracranial transplantation into individual embryos. This approach allows for the transplantation of hundreds of embryos per hour, thereby offering a scalable and orthotopic vertebrate DMG model that is directly applicable to high-throughput drug and gene discovery screens, which have the potential to accelerate clinical translation.

Key words Diffuse midline glioma, Zebrafish xenograft, Microglia, Confocal microscopy, Microplate reader

Michael Goldstein (ed.), *Translational Research Methods for Pediatric Brain Tumors*, Neuromethods, vol. 226,
https://doi.org/10.1007/978-1-0716-4754-7_13,

1 Introduction

Diffuse midline gliomas (DMGs) are rare central nervous system (CNS) tumors that originate in midline structures, including the thalamus, brainstem, and spinal cord [1, 2]. Classified as high-grade gliomas, DMGs are aggressive malignant tumors with a poor prognosis. They primarily affect children and young adults, with peak incidence observed in children aged 5–10 years [3, 4]. DMGs have a median survival of less than 1 year from diagnosis [5]. DMGs are highly metastatic tumors that infiltrate surrounding brain tissue, rendering surgical removal nearly impossible [6, 7]. Additionally, DMGs exhibit resistance to traditional chemotherapy and radiation therapy, which are standard treatments for other types of brain tumors [8]. Given the challenges in treating DMGs, understanding their complex biology within a natural physiological environment is essential for developing new therapeutic strategies. Enhancing the current animal models is a critical step toward achieving this goal [9].

Xenograft models, which involve the orthotopic transplantation of either patient-derived diffuse midline glioma cells or fresh brain tumor biopsy spheroids into immunodeficient rodents, closely mimic the genetic and phenotypic heterogeneity of the original patient tumor [10]. Due to their clinical relevance, xenograft models are considered an important model for studying the biology of DMG. However, despite their ability to sufficiently replicate human diffuse midline gliomas, orthotopic xenografts in rodents present significant technical challenges. The diffusely infiltrative nature of these gliomas creates, on average, a 6-month latency between tumor implantation and the death of mice from tumor growth [11]. While this long period of tumor growth and infiltration phenocopies the primary human tumor, the delay in reaching endpoints, such as survival or bioluminescent imaging of tumor growth, increases the time and expense of preclinical experimentation. Additionally, ethical considerations and high costs render DMG orthotopic xenografts impractical for large-scale screening and profiling efforts.

Over the past decade, the zebrafish (*Danio rerio*) has emerged as a powerful and clinically relevant model for studying human diseases, including cancer [12–17]. Zebrafish afford several key advantages, such as rapid development, small size, high reproductive capacity, and low-cost maintenance. Moreover, their lack of a functional adaptive immune system until 21 days postfertilization makes them particularly well-suited for xenotransplantation experiments [18]. Furthermore, zebrafish are exceptionally well-suited for large-scale drug discovery efforts, with their predictive capabilities already being utilized in numerous high-throughput screening studies [17, 19]. Recent studies have emphasized their remarkable ability to recapitulate disease characteristics and their strong clinical relevance [16, 20]. However, the intracranial transplantation of DMG cells into individual zebrafish embryos is technically demanding and time-consuming, which currently limits the use of zebrafish for studying the complex biology of DMG and conducting high-throughput screens for potential therapies. A novel method enabling the easy and rapid injection of DMG cells into zebrafish embryos would significantly advance our understanding of DMG biology and facilitate the development, characterization, and validation of new therapeutic options.

In this study, our objective is to establish a new zebrafish xenograft model that facilitates an easy and fast injection of DMG cells into zebrafish embryos. This method was adapted from established glioblastoma (GBM) xenotransplantation protocols to optimize the engraftment and survival of DMG cells in the zebrafish model [21]. To begin, we collected zebrafish embryos at ~1000-cell stage for microinjection, using both wild-type and transgenic animals (leptomeninges-YFP zebrafish and microglia-YFP zebrafish). These models are generated by integrating transgenic DNA fragments into the zebrafish genome, enabling stable overexpression of fluorescent proteins in specific cell compartments, such as microglial cells or leptomeningeal cells lining the ventricles. This approach facilitates visual confirmation of DMG localization and the investigation of interactions between DMG cells and microglia, providing valuable insights into the tumor microenvironment. At 3–4 days postinjection (dpi), confocal microscopy was employed to monitor tumor progression and assess the interactions between DMG cells and leptomeningeal cells or microglial cells of the injected embryos. A fluorescence microplate reader assay was used to quantify the RFP signal intensity [22, 23], providing insights into tumor cell proliferation and treatment responses (*see* Fig. 1). This approach has been used in other disease contexts for large-scale drug and gene screening in zebrafish larvae [24–27].

2 Materials

2.1 Cell Lines and Cell Culture

Cell lines: JHH-DIPG-2JA, JHH-DIPG-17, HSJD-007.

JHH-DIPG-2JA was developed from a posttreatment metastasis. JHH-DIPG-17 was developed from a pretreatment biopsy sample. HSJD-007 was developed from an autopsy sample at Hospital St. Joan de Deu in Barcelona, Spain, and was provided by the kind gift of Angel Carcaboso. Cell lines were confirmed to harbor the H3K27M mutation and were validated by STR testing. DMG cells were cultured in DMEM/F12 DIPG proliferation medium (EF medium), prepared to a final composition of: 30% Ham's F12 (Invitrogen, Cat# 11765-062), 70% DMEM (Invitrogen, Cat# 11965-118), 1% L-glutamine, 2% B27 supplement (Invitrogen, Cat# 12587-10), 20 ng/mL EGF (Peprotech, Cat# AF-100-15), 20 ng/mL FGF (Peprotech, Cat# 100-18B), and 5 μg/mL heparin.

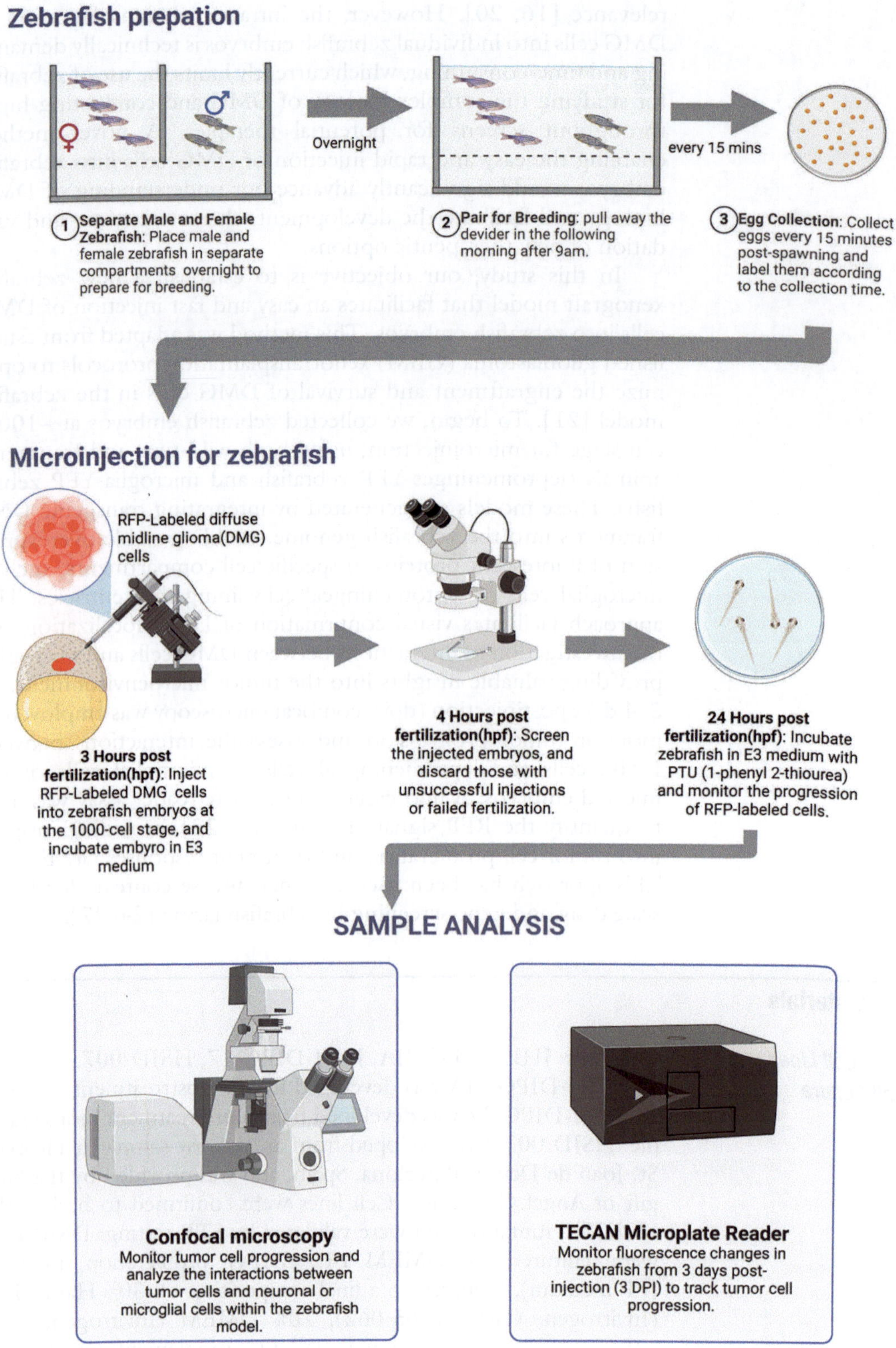

Fig. 1 Illustrations of the zebrafish embryo preparation, microinjection, and sample analysis

2.2 Lentiviral Transduction

For transduction, RFP-labeled lentivirus (Firefly Luciferase-T2A-RFP-IRES-Bsd Lentivirus, Biosettia, Cat# Glowcell-15b-1) was used. Blastcidin (Thermo Fisher Scientific, Cat# R21002), Polybrene (Millipore Sigma, Cat# H9268), and Accutase (Innovative Cell Technologies, Cat# A694-100ML) were also utilized in the transduction protocol.

2.3 Transgenic Zebrafish

To create "Microglia-YFP" zebrafish, a macrophage-specific Gal4 driver line [*Tg(mpeg1:GAL4-VP16)gl24*] was crossed to a UAS reporter-effector line co-expressing YFP and a nitroreductase 2.0 (NTR 2.0) [*Tg(5xUAS:GAP-TagYFP-P2A-NfsB_Vv_F70A/F108Y)jh513*]. To create "Leptomeninges-YFP" zebrafish, a leptomeninges-specific Gal4 driver line [*Tg(epd:Gal4)*] was crossed to a different UAS reporter-effector line co-expressing YFP and nitroreductase 2.0 (NTR 2.0) [28]

Tg(5xUAS:GAP-TagYFP-P2A-NfsB_Vv_F70A/F108Y, he:tag-BFP2)jh552.

2.4 Animal Studies

All procedures were conducted in compliance with the guidelines of the Office of Laboratory Animal Welfare (OLAW) for zebrafish research and under an approved protocol by the Johns Hopkins University Animal Care and Use Committee. Zebrafish were maintained under standard laboratory conditions at 28.5 °C with a 14-h light and 10-h dark cycle. The larvae designated for imaging were treated with 1-phenyl 2-thiourea (PTU) starting at 1 day postfertilization (dpf) to inhibit pigmentation.

2.5 Equipment and Materials

The following equipment and materials were used in this study:

1. Confocal Microscope: Olympus FV1000 equipped with 515 nm and 559 nm laser lines. Stacked confocal images were captured using an XLUMPLFL 20X W NA: 0.95 objective, with a step size of 5 μm.
2. Fluorescence Microscope: Motic AE31.

   Fluorescence microplate reader (TECAN Infinite M1000 PRO; excitation 555 nm, bandwidth 5 nm; emission 585 nm, bandwidth 10 nm).
3. Microinjection System: NARISHIGE IM-200 Microinjector.
4. Needle Puller: Sutter Instrument Model P-97, configured with the following settings: Heat 580, Pull 70, velocity 90, Time 110.
5. Glass Capillaries: 4-inch Borosilicate Glass, 1.2 mm OD (Cat# 1B120-4, WPI).
6. Microloader Tips: Caliber: 2 × 96 (Cat# EPE-930001007).
7. Microplate: 96-well, polypropylene (PP), U-bottom, black (Greiner Bio-One, Cat# 650209).

2.6 Other Reagents

1. Tricaine methanesulfonate anesthetic (Fisher, NC0342409, final concentration: 0.612 nM).
2. Low-melt agarose (Fisher, BP1360-100).
3. E3 media: Prepare 1 L of 100X E3 medium by dissolving 29.22 g of NaCl, 1.27 g of KCl, 3.33 g of $CaCl_2$, and 3.97 g of $MgSO_4$ in water. Adjust the pH to 7.4 using 5 M NaOH. To obtain 1× E3 medium, dilute the 100× E3 stock solution with water. The final solute concentrations in 1× E3 medium are NaCl (5 mM), KCl (0.17 mM), $CaCl_2$ (0.3 mM), and $MgSO_4$ (0.33 mM).
4. PTU pigmentation inhibition agent (1-phenyl-2-thiourea, Fisher-P023725G, final concentration: 200 nM).

3 Methods

3.1 Lentivirus Transduction

The lentivirus used contains a blasticidin resistance gene for selection. Prior to transduction, cell lines were tested for blasticidin tolerance to determine the optimal concentration. Cells were seeded in a 24-well plate, with each well containing the appropriate number of cells (e.g., 20,000 cells per well for HSJD007). Eleven blasticidin concentration groups were prepared, each treatment was performed in duplicate wells, with concentrations ranging from 0 to 10 μg/mL. Blasticidin was added, and cells were monitored for 1 week. Cell health was assessed daily under a microscope, and cell viability was confirmed using CellTiter-Blue. The lowest blasticidin concentration that effectively killed all cells was determined for use in subsequent experiments.

Lentivirus (Firefly Luciferase-T2A-RFP-IRES-Bsd Lentivirus, Biosettia, Cat# Glowcell-15b-1) was added to the cells based on the cell density in each dish. Depending on the cell line, 5–10 μg/mL of polybrene (Millipore Sigma, Cat# H9268) was used. After 24 h of incubation, cells were centrifuged, and the medium was replaced with fresh culture medium. Two days post-transduction, cells were cultured in medium with the predetermined concentration of blasticidin for 1 week, with regular medium changes. Transduction efficiency was quantified by determining the percentage of RFP-positive DMG cells using fluorescence microscopy. And, for patient-derived DMG cells, single-cell sorting is not recommended, as it may result in the loss of many cells due to their large size and damage during dissociation and sorting.

3.2 Cell Preparation

3.2.1 Thawing DMG Cell Lines (JHH-DIPG-17, JHH-DIPG-2JA, HSJD-007)

Cryopreserved DMG cell lines were thawed by removing vials from liquid nitrogen storage and placing them directly into a 37 °C water bath. Once thawed, cells were transferred gently using a 1 mL pipette, Pasteur pipette, or P1000 pipette tip into a 15 mL conical tube. Pre-warmed medium was added dropwise (~200 μL per drop) using a P1000 pipette, with gentle swirling after each drop to gradually equilibrate cells from their cryoprotectant solution to isotonic conditions. This process continued until a final volume of 4 mL was reached. Cells were then centrifuged, the supernatant was removed, and the cell pellet was resuspended in 10 mL of fresh DMEM/F12 DIPG proliferation medium (EF medium). Cells were seeded into T25 flasks with regular medium change.

3.2.2 Cell Collection and Preparation for Zebrafish Embryo Injection

To prepare cells for injection into zebrafish embryos, the cell suspension was transferred to a 15 or 50 mL conical tube and centrifuged to collect the cell pellet. Given the tendency of DMG cells to form tumor spheroids, the supernatant was carefully aspirated, and Accutase was added to the cell pellet to promote dissociation into single cells. The suspension was then gently mixed and incubated in a 37 °C water bath for 5 min. Following incubation, cells were centrifuged at 1000×*g* for 5 min. The Accutase was aspirated, and the cells were resuspended by pipetting with a 200 μL tip to ensure a single-cell suspension. The suspension was then passed through a 40 μm nylon mesh to further refine the single-cell preparation [21].

Cell counts were performed using a hemocytometer under a fluorescence microscope to quantify both the total cell count and the proportion of RFP-labeled cells. The cell suspension was washed three times with ice-cold phenol red-free HBSS buffer to eliminate any residual growth factors or medium components that could be toxic to zebrafish embryos. Finally, cells were resuspended in ice-cold HBSS medium containing 2% polyvinylpyrrolidone (Sigma) to avoid clogging of the microinjection capillary at a concentration of 100 cells/nL [21].

3.3 Fish Preparation

Zebrafish breeding pairs were set up in the afternoon (the day before the injection) to ensure fertilized embryos were available on the injection day. The setup, illustrated in Fig. 2a, includes the tank, breeding divider, inserted breeding ramp, and lid arranged from left to right. The tank was assembled with the divider positioned to separate male and female zebrafish.

On the following morning, the divider was removed to allow breeding, and embryos were collected at 15-min intervals. Each collection was labeled with the corresponding time of collection. The collected embryos were placed in a 28.5 °C incubator. After 3 h of incubation, the embryos were examined under a microscope to select ~1000-cell stage embryos for microinjection.

3.4 Fish Injection

When the zebrafish embryos were close to ~1000-cell stage, tumor cells were prepared for injection. Embryos were aligned in the mold (*see* Fig. 3c) and oriented using a pipette tip so that the blastoderm was facing upwards, facilitating injection and minimizing injury (*see* Fig. 3d).

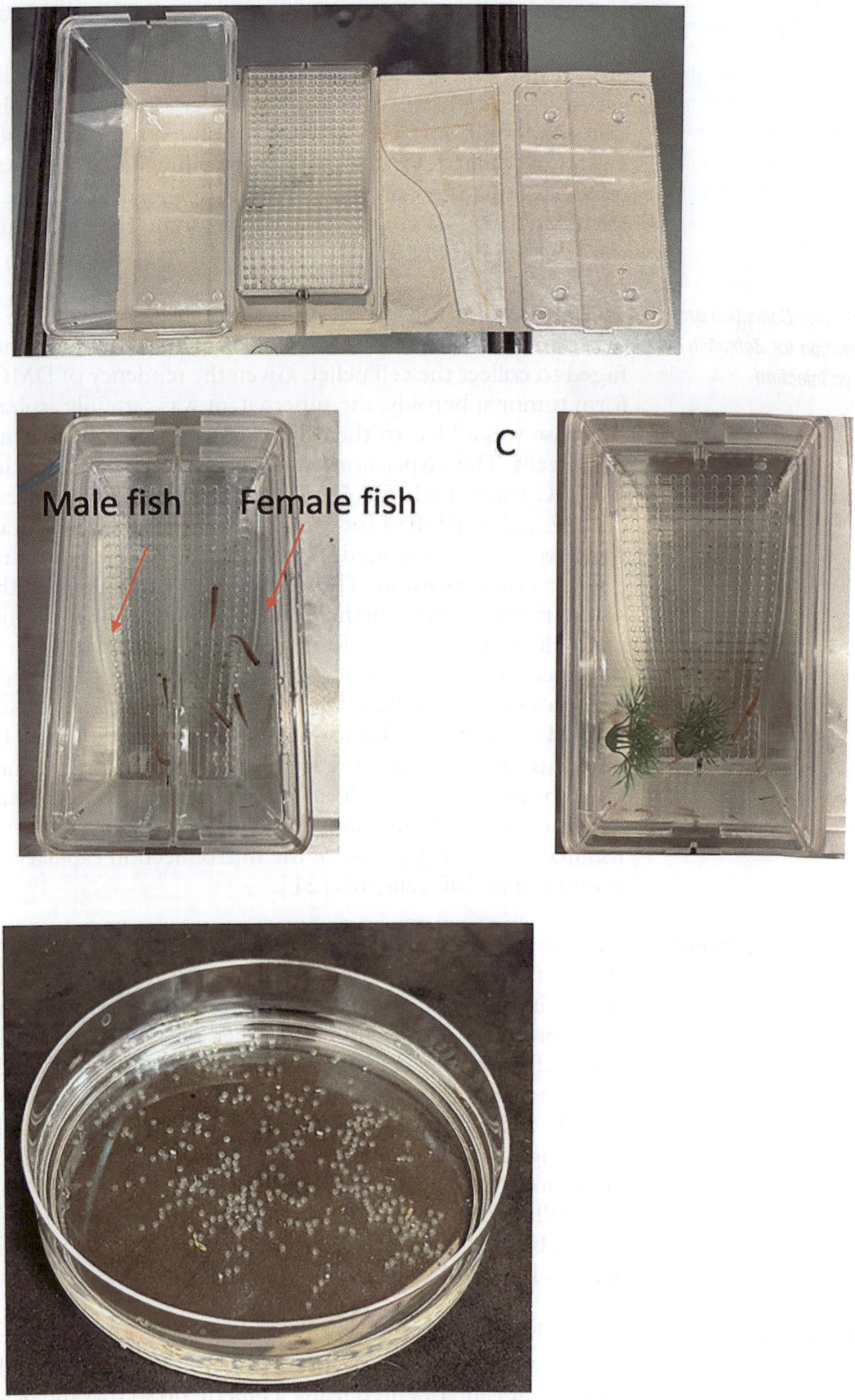

Fig. 2 Breeding setup and procedure for zebrafish. (**a**) Components of the breeding setup, including the tank, breeding ramp, dividers, and lid (from left to right). (**b**) Zebrafish tank setup for overnight breeding. (**c**) Breeding tank with the slide removed to initiate spawning. (**d**) Collected embryos. Embryos are collected every 15 min and labeled with the corresponding collection time

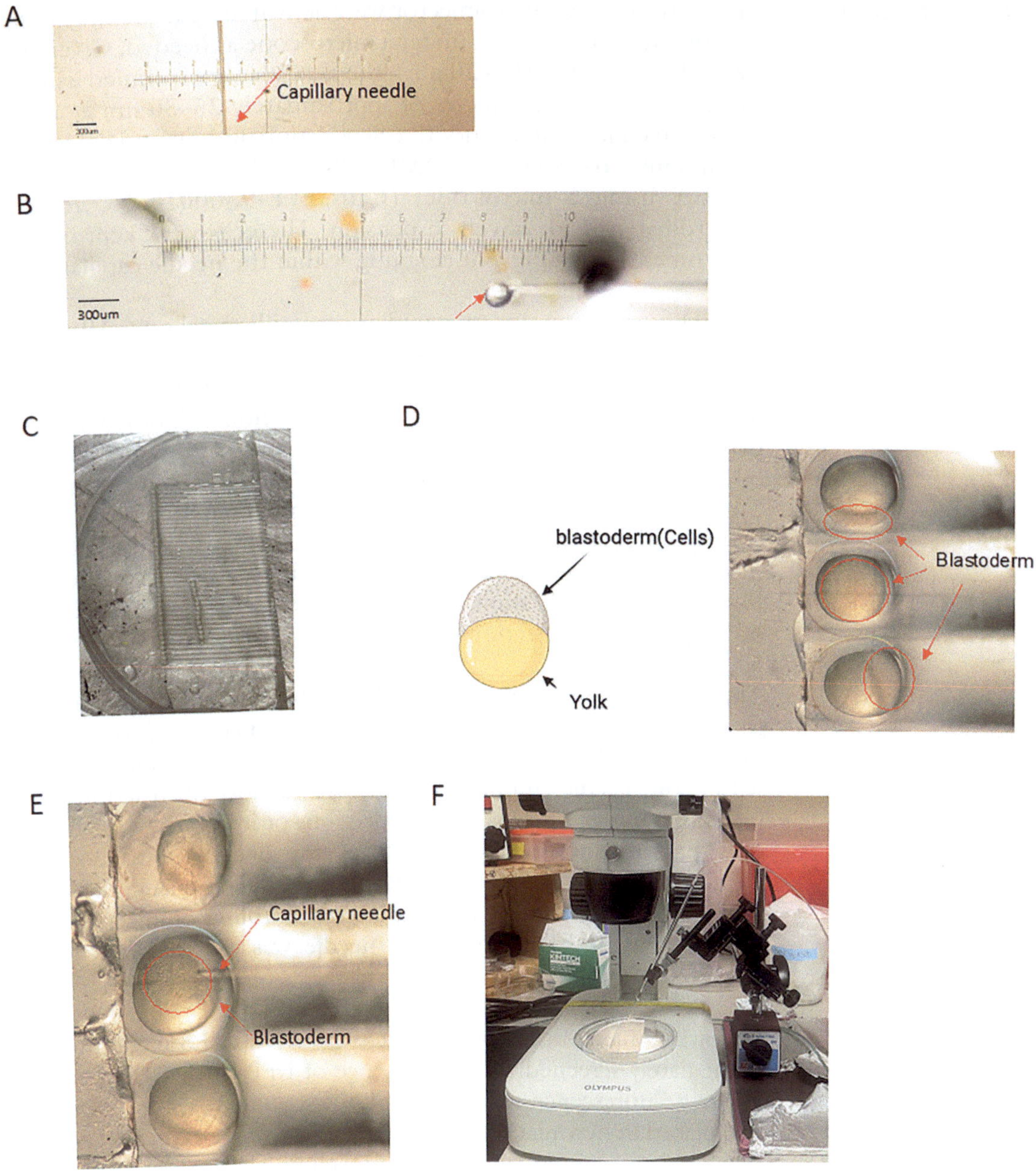

Fig. 3 Microinjection procedure and apparatus for zebrafish embryos. (**a**) Microinjection capillary used for injections, with ruler under the microscope. Each tick mark represents 30 μm, with the needle trimmed to a diameter of approximately 20–25 μm. (**b**) Measurement of the injection droplet size (with a diameter of approximately 130 μm and corresponding to a volume of approximately 1 nL) under the microscope to ensure accurate volume delivery. Approximately 100 tumor cells are injected into each embryo. (**c**) Mold used to hold embryos during microinjection. (**d**) The left cartoon illustrates the blastoderm and yolk of a zebrafish embryo at the 1000-cell stage. The right image shows the orientation of a 1000-cell stage embryo for injection, with the blastoderm positioned upward. (**e**) The injection of embryos undergoing microinjection. (**f**) Apparatus showing the microinjection holder and the needle angle (approximately 60°)

3.4.1 Injection Setup

The air source and microinjector were activated, and the tips of the injection needles were cut under a microscope as needed. Given the size of human cells (10 ~ 20 μm), the needles were trimmed using a sharp blade to a tip diameter of approximately 25–30 μm at a 45° angle, which facilitated smooth cell injection and minimized injury for the embryo (*see* Fig. 3a). DMG cells were loaded into the needle capillary using a microloader (Cat# EPE-930001007), adding approximately 4 μL of cell suspension. The cells were kept on ice at all times to help prevent clogging within the needle capillary.

3.4.2 Injection Calibration

The cell suspension was prepared at one million cells per 10 μL to achieve the desired density (100 cells/nL). The needle was back-filled with DMG cells and securely mounted in the microinjector holder of the microinjection system. Injection solution was dispensed using a foot pedal. The needle tip was placed into mineral oil, and DMG cells were injected into the oil to produce droplets. The droplet size, with a diameter of approximately 130 μm and corresponding to a volume of approximately 1 nL, was adjusted and measured by fine-tuning the injection time and air pressure (*see* Fig. 3b). After calibration, the capillary needle was positioned at a 60° angle for the injection procedure.

3.4.3 Embryo Injection Procedure

The embryos and needle were brought into focus under the microscope, ensuring that the injection angle remained at approximately 60° (*see* Fig. 3f). This angle helps prevent cell clumping of cells within the needle, minimizing the risk of flow obstruction and injection failures, such as empty injections or injections containing only buffer. Additionally, it reduces the likelihood of needle-induced injury to the embryo. During the experiment, mineral oil was used to check needle patency after each round of injections. If cell clumping was observed, the needle was replaced to ensure consistent performance.

3.4.4 Postinjection Embryo Care

Following injection, embryos were transferred to E3 medium and immediately screened for the presence of RFP-labeled DMG cells under a fluorescence microscope. Positive embryos were selected and placed in new plates, ensuring adequate space to prevent overcrowding and subsequent embryo mortality. The morning after the injection, the E3 medium was replaced with E3 medium supplemented with 1-phenyl 2-thiourea (PTU) to inhibit pigmentation, thereby facilitating live imaging. Embyros were incubated in 28.5 °C degree incubator. Embryos were monitored daily, and dead embryos were removed promptly to prevent contamination. Fresh E3 + PTU medium was changed every other day. At 5 days postinjection, zebrafish were fed paramecia to facilitate larval development.

3.5 Tracking Tumor Cell Progression in Zebrafish

Before the time-course imaging, the fish were anesthetized by transferring them into embryo medium supplemented with tricaine, allowing approximately 3 min for them to become unresponsive to touch. Subsequently, they were transferred to a transparent 96-well plate. A homemade hair loop or 10 μL pipette tip was employed to position and orient the fish appropriately. Time-course images were captured using fluorescence microscopy at the initial time point (0 h), and subsequently at 24, 48, and 72 h postinjection to monitor the migration of tumor cells (*see* Fig. 5). After imaging, the fish were thoroughly washed to remove tricaine and placed back into E3 + PTU medium for incubating until the following imaging series. The time-course images (Fig. 4a, b) display an example of blastulas injected with JHH-DIPG-2JA cells or JHH-DIPG-17 cells. These images demonstrate the proliferation and migration of DMG cells inside the zebrafish.

3.6 Quantifying RFP-Labeled DMG Cell Fluorescence in Zebrafish

Fluorescence intensity was quantified using a TECAN microplate reader at 3 and/or 4dpi. Individual zebrafish were anesthetized with tricaine and placed in separate wells of a black 96-well plate (Fig. 5a). Non-injected zebrafish were included for defining the fluorescent signal cutoff. To ensure accuracy and minimize variation in the microplate reader results, the zebrafish larvae were washed multiple times with clean PTU/E3 medium to remove any variations caused by impurities in the water.

All injected zebrafish displaying RFP fluorescence at 3 days postinjection were selected and washed several times with clean E3 medium. Tricaine (working concentration: 0.612 nM) was then added, and the zebrafish were individually transferred to a well of a black 96-well plate for further analysis. Each zebrafish was positioned to make sure its head was centered within the well under microscope. The plate was then loaded into the TECAN microplate reader to measure RFP fluorescence intensity (excitation 555 nm, emission 585 nm). The TECAN microplate reader was configured to capture multiple data points per each well to ensure comprehensive coverage of the entire well area (*see* Fig. 5b); see White et al., for full methodological details [23].

Briefly, non-injected zebrafish were used to establish a fluorescent signal cutoff value defined as the average of the maximal fluorescence intensity value for each non-injected zebrafish plus three standard deviations (avg + 3SD; *see* Fig. 5c). Total fluorescent signal for tumor-injected zebrafish was calculated by summing all regional scans greater than the signal cutoff for each individual sample (*see* Fig. 5c). Fluorescence readings below the signal cutoff were considered regions of the well that did not contain a fish and were thus excluded from the analysis. Fluorescence intensity was measured daily for each injected zebrafish to track changes over time. Figure 6b illustrates the relative fluorescence changes in zebrafish from Day 4 postinjection, while shows fluorescence intensity persists over time.

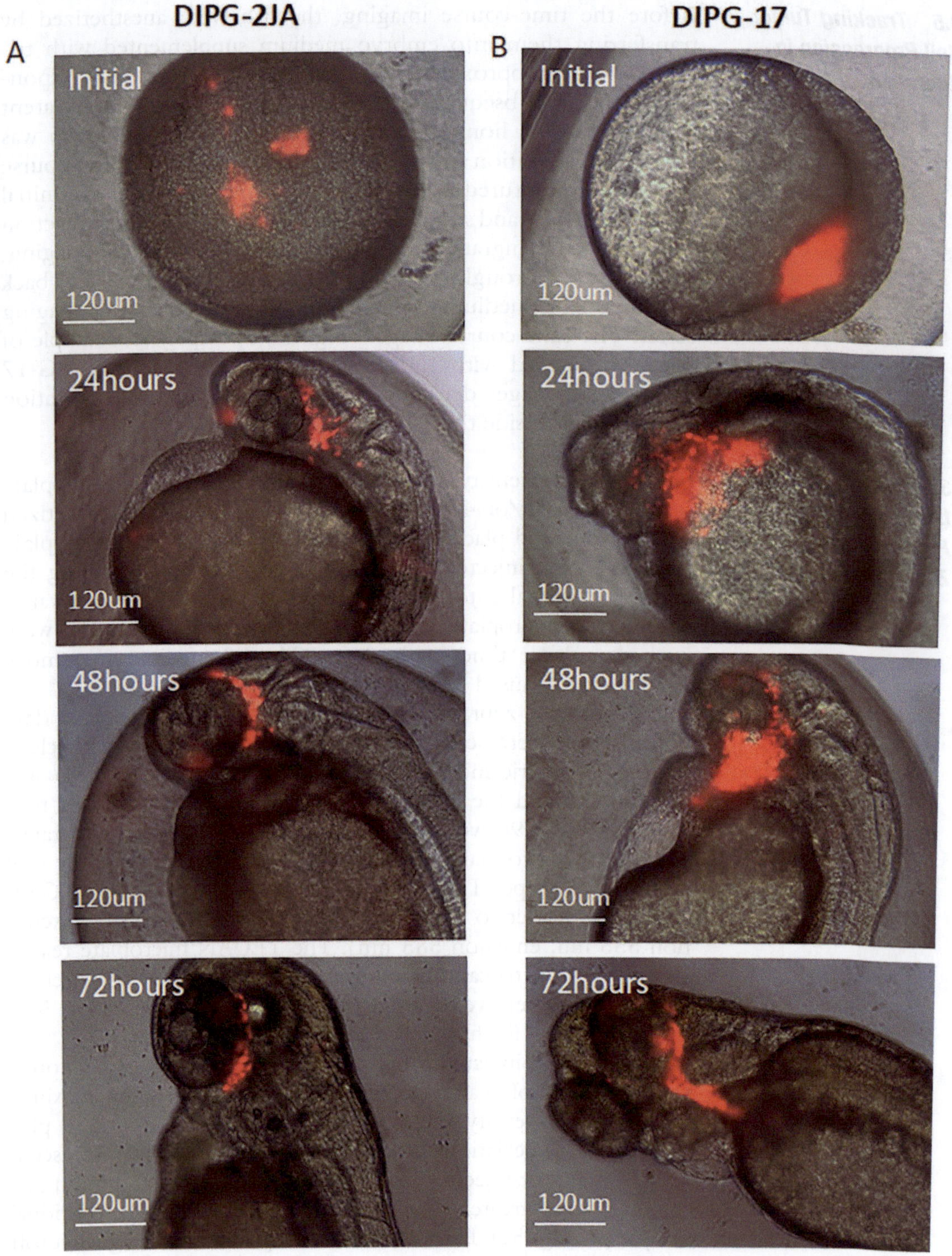

Fig. 4 Time-course images after blastula injections. (**a**) Time-course images showing the progression of DIPG-2JA cells following blastula-stage injection in zebrafish embryos. Fluorescent images were captured at the initial injection (0 h), and at 24, 48, and 72 h post-injection, illustrating the distribution of RFP-labeled tumor cells over time. Scale bar: 120 μm. (**b**) Time-course images following blastula-stage injection of the DIPG-17 cell line in zebrafish embryos. Images were captured at the initial injection (0 h) and at 24, 48, and 72 h post-injection, showing the distribution of RFP-labeled DIPG-17 tumor cells over time. Scale bar: 120 μm

A

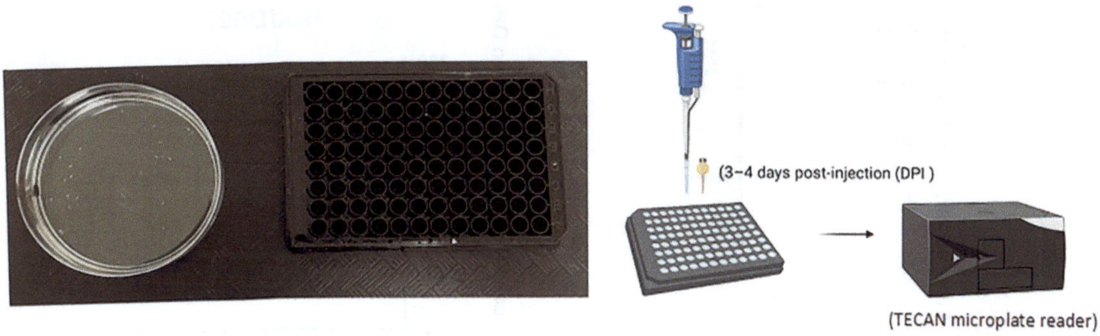

B

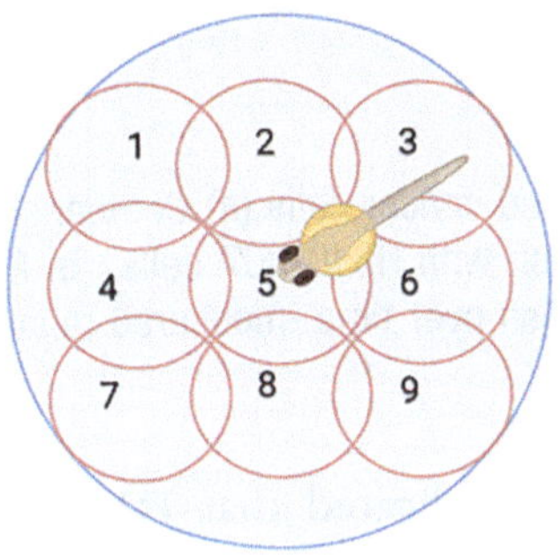

C

Fluorescent signal cutoff value =Average+3SD

SD = standard deviation of the maximal regional value observed across all non-injected controls

Average= average of the maximal regional value observed across all non-injected controls

Fluorescent signal for tumor-injected zebrafish = Sum of regional scans greater than the signal cutoff

Fig. 5 The setup and analysis of zebrafish fluorescence intensity by TECAN microplate reader. (**a**) The left image shows a black U-shape 96-well plate and zebrafish larvae. The right image illustrates the procedure of placing zebrafish into the 96 well followed by microplate reading. (**b**) Alignment of the nine measurement points per well, representing the reading grid within the well. (**c**) Non-injected zebrafish were used to establish a fluorescent signal cutoff value defined as the average of the maximal fluorescence intensity value for each non-injected zebrafish plus three standard deviations (avg + 3SD). Total fluorescent signal for tumor-injected zebrafish was calculated by summing all regional scans greater than the signal cutoff for each individual sample

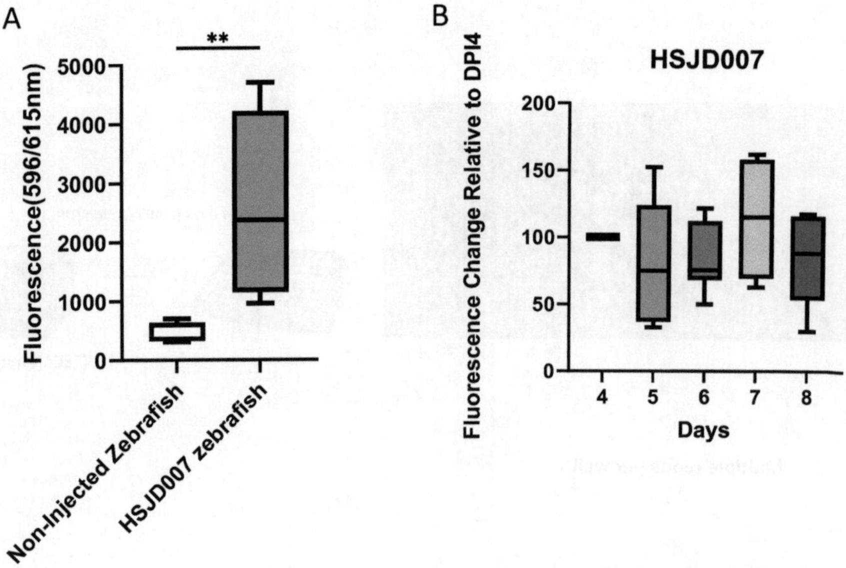

Fig. 6 Fluorescence analysis of zebrafish injected with RFP-labeled cancer cells (**a**) Comparison of fluorescence intensity between non-injected and those injected zebrafish with HSJD-007 cells. (**b**) Mean fluorescence intensity change of RFP-labeled HSJD-007 cells in zebrafish over time, measured relative to 4 days postinjection (4 dpi) and monitored through 8 dpi

3.7 Confirmation of DMG Cells in the Brain in Zebrafish

Although time-course imaging indicated that DMG cells at the 1000-cell stage injected into zebrafish embryos migrated toward the head, it remained unclear if these cells specifically migrated into the brain. To investigate this, DMG cells were injected into transgenic zebrafish embryos expressing YFP-tagged leptomeningeal cells or microglia/macrophages (see Methods for transgene details). At 6 days postinjection (dpi), confocal live imaging was performed to access DMG localization.

Preparation for the whole fish imaging: Several hours before imaging, we dissolved low melt agarose (Fisher, BP1360-100) in PTU/E3 medium to achieve a final concentration of 1% and maintained it at 40 °C. At least 30 min before imaging, we prepared the mounting solution by combining 1% LMA with tricaine, mixed thoroughly, and returned the solution to 40 °C. We anesthetized the fish by immersing them in Tricaine/PTU/E3 and waited approximately 3 min for them to become unresponsive to touch. Subsequently, we transferred individual fish into mounting medium and then positioned them within a drop (25–30 μL) of the medium on a 10 cm petri dish for imaging using an upright confocal microscope. A homemade hair loop was used to position the fish. When the agarose became solidified, the plate was moved to the confocal stage. Then, 50 mL of Tricaine/PTU/E3 was added to the petri dish for imaging using water immersion objectives. Z-stacks were obtained at a 5 μm z-depth resolution [29].

Next, ".Oib" files for each image were loaded into IMARIS (v9.7.0; Bitplane) for 3D rendering. Figure 7c demonstrates

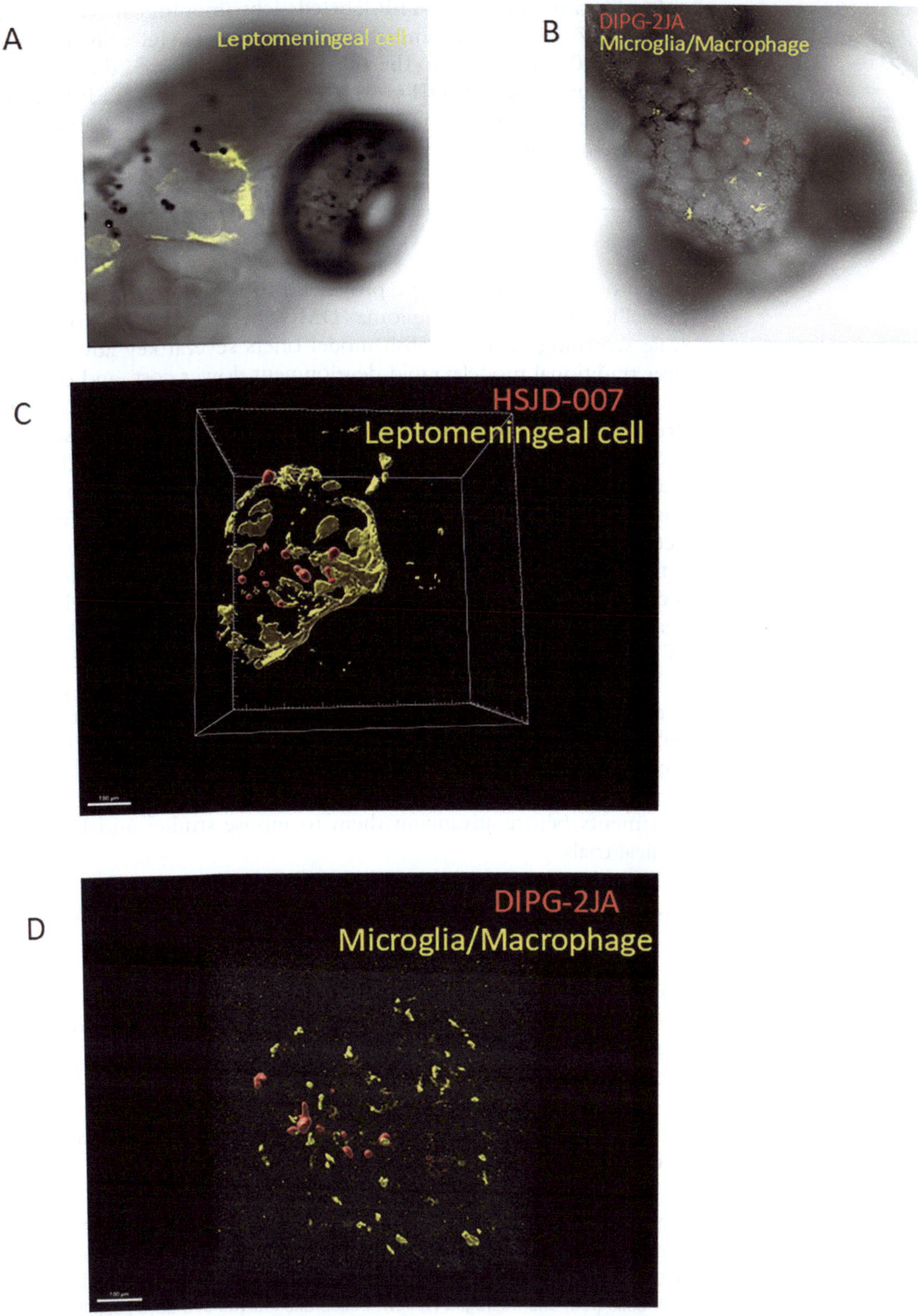

Fig. 7 Colocalization analysis of YFP-labeled brain leptomeningeal cells and microglial cells with RFP-labeled DMG cells. (**a**) Confocal image of a zebrafish head showing the position of yellow fluorescent protein (YFP)-labeled leptomeningeal cells. (**b**) Confocal image of a zebrafish head displaying the position of YFP-labeled

colocalization between YFP-labeled leptomeningeal cells and HSJD-007 cells, confirming the migration of DMG cells injected at the 1000-cell stage into the zebrafish brain. Figure 7d illustrates the colocalization of DIPG-2JA (RFP) cells with YFP-labeled microglia cells, showing interactions between DMG cells and the microenvironment.

4 Notes

The procedure outlined here provides a robust zebrafish xenograft model for diffuse midline glioma (DMG) that is amenable to large-scale screening. The zebrafish model offers several key advantages over traditional models: rapid development, lower cost, and optical transparency, making it an efficient platform for high-throughput drug screening. By monitoring the migration of fluorescence protein-positive DMG cells, researchers can track tumor burden in real time in a vertebrate model. The use of confocal microscopy enables the acquisition of time-course z-stack images, facilitating detailed assessments of tumor cell migration, proliferation, and interactions with the microenvironment. This approach significantly reduces the time and financial investment required for preclinical testing of new drugs that affect DMG viability, mobility, or the ability of DMG cells to interact with the immune microenvironment.

This model thus serves as a valuable tool for both high-throughput drug discovery and fundamental research on tumor biology. In summary, our zebrafish DMG xenograft model offers a cost-effective, scalable, and versatile platform for preclinical drug evaluation, potentially accelerating the identification of effective treatments before advancing them to mouse studies and human clinical trials.

Acknowledgments

Funding for this research was provided by The Cure Starts Now Foundation, The Giant Food Pediatric Cancer Research Fund, and the NCI Core Grant to the Johns Hopkins Sidney Kimmel Comprehensive Cancer Center (P30CA006973).

Conflict of Interest The authors have no conflicts of interest to declare that are relevant to the content of this chapter.

Fig. 7 (continued) microglia and DIPG-2JA cells. (**c**) Colocalization of YFP-labeled brain leptomeningeal cell network (yellow) and RFP-labeled HSJD-007 tumor cells (red) in zebrafish at 6 days postinjection (dpi). (**b**) Colocalization of DIPG-2JA tumor cells (red) and microglia (yellow) in zebrafish larvae at 6 days postinjection (dpi). Confocal fluorescence imaging reveals the presence of DMG cells within the brain parenchyma and spatial proximity between tumor cells and microglia/macrophages

References

1. Hoffman LM, Veldhuijzen van Zanten SEM, Colditz N, Baugh J, Chaney B, Hoffmann M et al (2018) Clinical, radiologic, pathologic, and molecular characteristics of long-term survivors of diffuse intrinsic pontine glioma (DIPG): a collaborative report from the International and European Society for Pediatric Oncology DIPG Registries. J Clin Oncol 36: 1963–1972
2. Ostrom QT, Cioffi G, Gittleman H, Patil N, Waite K, Kruchko C et al (2019) CBTRUS statistical report: primary brain and other central nervous system tumors diagnosed in the United States in 2012-2016. Neuro-Oncology 21:v1–v100
3. Filbin M, Monje M (2019) Developmental origins and emerging therapeutic opportunities for childhood cancer. Nat Med 25:367–376
4. Rashed WM, Maher E, Adel M, Saber O, Zaghloul MS (2019) Pediatric diffuse intrinsic pontine glioma: where do we stand? Cancer Metastasis Rev 38:759–770
5. Louis DN, Giannini C, Capper D, Paulus W, Figarella-Branger D, Lopes MB et al (2018) cIMPACT-NOW update 2: diagnostic clarifications for diffuse midline glioma, H3 K27M-mutant and diffuse astrocytoma/anaplastic astrocytoma, IDH-mutant. Acta Neuropathol (Berl) 135:639–642
6. Di Ruscio V, Del Baldo G, Fabozzi F, Vinci M, Cacchione A, de Billy E et al (2022) Pediatric diffuse midline gliomas: an unfinished puzzle. Diagnostics 12:2064
7. Schwartzentruber J, Korshunov A, Liu X-Y, Jones DTW, Pfaff E, Jacob K et al (2012) Driver mutations in histone H3.3 and chromatin remodelling genes in paediatric glioblastoma. Nature 482:226–231
8. Buczkowicz P, Hawkins C (2015) Pathology, molecular genetics, and epigenetics of diffuse intrinsic pontine glioma. Front Oncol [Internet] [cited 2024 Nov 19];5. Available from: https://www.frontiersin.org/journals/oncology/articles/10.3389/fonc.2015.00147/full
9. Kegelman TP, Hu B, Emdad L, Das SK, Sarkar D, Fisher PB (2014). *In vivo* modeling of malignant glioma. In: Tew KD, Fisher PB (eds) Advances in cancer research [Internet]. Academic Press [cited 2024 Nov 19]. pp 261–330. Available from: https://www.sciencedirect.com/science/article/pii/B978012800249000007X
10. Mann B, Artz N, Darawsheh R, Kram DE, Hingtgen S, Satterlee AB (2025) Opportunities and challenges for patient-derived models of brain tumors in functional precision medicine. NPJ Precis Oncol 9:1–16
11. Taylor IC, Hütt-Cabezas M, Brandt WD, Kambhampati M, Nazarian J, Chang HT et al (2015) Disrupting NOTCH slows diffuse intrinsic pontine glioma growth, enhances radiation sensitivity, and shows combinatorial efficacy with bromodomain inhibition. J Neuropathol Exp Neurol 74:778–790
12. Veinotte CJ, Dellaire G, Berman JN (2014) Hooking the big one: the potential of zebrafish xenotransplantation to reform cancer drug screening in the genomic era. Dis Model Mech 7:745–754
13. Zebrafish cancer: the state of the art and the path forward. Nat Rev Cancer [Internet]. [cited 2024 Nov 19]. Available from: https://www.nature.com/articles/nrc3589
14. Lieschke GJ, Currie PD (2007) Animal models of human disease: zebrafish swim into view. Nat Rev Genet 8:353–367
15. Zebrafish in vivo models of cancer and metastasis – PMC [Internet]. [cited 2025 Feb 13]. Available from: https://pmc.ncbi.nlm.nih.gov/articles/PMC7397842/
16. Patton EE, Zon LI, Langenau DM (2021) Zebrafish disease models in drug discovery: from preclinical modelling to clinical trials. Nat Rev Drug Discov 20:611–628
17. Goldrick C, Palanga L, Tang B, Mealy G, Crown J, Horgan N et al (2021) Hindsight: review of preclinical disease models for the development of new treatments for uveal melanoma. J Cancer 12:4672–4685
18. Traver D, Herbomel P, Patton EE, Murphey RD, Yoder JA, Litman GW et al (2003) The zebrafish as a model organism to study development of the immune system. Adv Immunol 81:253–330
19. MacRae CA, Peterson RT (2015) Zebrafish as tools for drug discovery. Nat Rev Drug Discov 14:721–731
20. Pliakopanou A, Antonopoulos I, Darzenta N, Serifi I, Simos YV, Katsenos AP et al (2023) Glioblastoma research on zebrafish xenograft models: a systematic review. Clin Transl Oncol 26:311
21. Pudelko L, Edwards S, Balan M, Nyqvist D, Al-Saadi J, Dittmer J et al (2018) An orthotopic glioblastoma animal model suitable for high-throughput screenings. Neuro-Oncology 20:1475–1484
22. Walker SL, Ariga J, Mathias JR, Coothankandaswamy V, Xie X, Distel M et al

(2012) Automated reporter quantification in vivo: high-throughput screening method for reporter-based assays in zebrafish. PLoS One 7:e29916

23. White DT, Eroglu AU, Wang G, Zhang L, Sengupta S, Ding D et al (2016) ARQiv-HTS, a versatile whole-organism screening platform enabling in vivo drug discovery at high-throughput rates. Nat Protoc 11:2432–2453
24. Emmerich K, Hageter J, Hoang T, Lyu P, Sharrock AV, Ceisel A et al (2024) A large-scale CRISPR screen reveals context-specific genetic regulation of retinal ganglion cell regeneration. Dev Camb Engl 151:dev202754
25. Zhang L, Chen C, Fu J, Lilley B, Berlinicke C, Hansen B et al (2021) Large-scale phenotypic drug screen identifies neuroprotectants in zebrafish and mouse models of retinitis pigmentosa. elife 10:e57245
26. Wang G, Rajpurohit SK, Delaspre F, Walker SL, White DT, Ceasrine A et al (2015) First quantitative high-throughput screen in zebrafish identifies novel pathways for increasing pancreatic β-cell mass. elife 4:e08261
27. Kelpsch DJ, Zhang L, Thierer JH, Koren K, Kumar U, Lin Y, et al (2025) A whole-animal phenotypic drug screen identifies suppressors of atherogenic lipoproteins. BioRxiv Prepr Serv Biol; 2024.11.14.623618
28. Sterrer S, Königstorfer A, Hoffman W (1990) Biosynthesis and expression of ependymin homologous sequences in zebrafish brain. Neuroscience 37:277–284
29. White DT, Mumm JS (2013) The nitroreductase system of inducible targeted ablation facilitates cell-specific regenerative studies in zebrafish. Methods 62:232–240

Index

B

C

D

E

G

I

L

M

N

O

P

R

S

Michael Goldstein (ed.), *Translational Research Methods for Pediatric Brain Tumors*, Neuromethods, vol. 226, https://doi.org/10.1007/978-1-0716-4754-7,

MIX
Papier aus verantwortungsvollen Quellen
Paper from responsible sources
FSC® C105338

If you have any concerns about our products,
you can contact us on
ProductSafety@springernature.com

In case Publisher is established outside the EU,
the EU authorized representative is:
Springer Nature Customer Service Center GmbH
Europaplatz 3, 69115 Heidelberg, Germany

Printed by Libri Plureos GmbH
in Hamburg, Germany